The Ultimate Guide to STARGAZING

Second Edition 2023

VOLUME 4

COSMOLOGY

The Ultimate Guide to Stargazing - Volume 4

Second Edition 2023

ISBN: 978-0-6454669-7-3 (paperback)

First published by Dreamtech Designs & Productions Pty Ltd 2018 ©

DEDICATION

I dedicate this book to my family. Firstly, to my wife Deborah for her loyal support and for the encouragement she has given me over the many years it has taken to do the research, produce the imagery and write *STARGAZING's* volumes. I have greatly appreciated her advice and discussions on subject matter and content as well as the time she invested in assisting with editing. I am also in debt to my very supportive son Tane for publishing and marketing this book, as well as his feedback on certain subjects and his help with editing over the years. If it was not for his dedication and his expertise in managing our tourist attractions, I would not have had the freedom or the time to do the massive amount of research required, or the time to write this book.

I must give a special thanks to my father and mother who both did so much to encourage my endeavors in astronomy and in the other sciences from when I was only eight. My father bought me expensive books on astronomy and the other sciences, as well as buying for me my first telescopes, a microscope, a chemistry set, electricity and magnetism experiments, gliders, model aeroplanes, and model boats, as well as a large trains set, and many puzzles – all of which I treasured. In my youth before I was old enough to drive, my wonderfully considerate mother would drive me and my telescope out to my uncle's farm in the country, so I could observe deep sky objects in their full glory.

Finally, I wish to thank my beautiful and intelligent granddaughters Aisja and Sienna, who have given me much inspiration and joy and who are the future of our family. They will live in the world of increasingly exciting inventions and discoveries, incredible concepts and extraordinary change, as explained in this book.

The Milky Way rising behind a quaint stone church on a hilltop in New Zealand photographed by Nico Babot.

CONTENTS

ACKNOWLEDGMENTS

I give special thanks to the following people who offered valuable suggestions and comments when reading through the entire manuscript. They were my wife, Deborah Kelly, Nick Parfitt, John O'Sullivan Greg Bock and James Bryan. Others graciously assisted with various chapters. They included my son, Tane Thompson, Olivia King, Terry Lovejoy, Robert Stanford, Joe Gjerek, Dan McDonald, Brandon Clift, and Charles Gower. Nicole Brooke and Rebecca Gjerek worked with me on producing many of the diagrams. Joe Gjerek designed the excellent volume covers based on my rough ideas and his wife Rebecca produced a number of the complex CGI illustrations under my direction.

I must thank the contribution that my high school physics teacher Mr Thompson (no relation) and my chemistry teacher Mr McAlpine made by instilling in me the importance of scientific methodology, and importantly, the thrill of discovery and how to go about making accurate observations and how to conduct useful experiments (which unfortunately, has now been largely removed from high school science by non-scientific public servants with the view of eliminating all risk). While my tertiary education was valuable, it did not spark as much excitement as those early years of discovering the 'Wow!' factor in scientific inquiry and observational astronomy. In my adult working years when I built my own businesses, I found that my self-education in the sciences to be very enlightening. I gained the ability to learn across many fields and to see the connections between all the sciences. This proved to be very valuable in writing the last volume of this book.

Over the decades since I was in secondary school, there have been many amateur and professional astronomers in many countries, too numerous to mention here, who have helped me learn about a very broad range of aspects of astronomy. Many supported my astronomical endeavors in various ways, so I thank them for the knowledge they gave me to allow me to write this book. My good friend and expert telescope maker Cliff Duncan taught me how to build telescopes, and he also built excellent telescopes for me to use. He challenged me to think deeply about the many subjects that we often discussed at length. There have been many amateur astronomers that I have enjoyed

the company of when at dark sky astrocamps where we would observe deep sky objects long into the night. There were also those who assisted me in creating and testing the Supernova Search Charts. Many professional astronomers believed in the value of those charts for making early discoveries of supernovas, and in doing so, they gave me much encouragement to complete the project, for which I thank them.

I have had a lifetime observing in the field as well as at private and professional observatories, and of course, in my own observatory. Over my life, I have had numerous discussions late into the night with both amateur and professional astronomers who gave me a diversity of insights into many aspects of astronomy and science in general. With their knowledge and my love of all aspects of astronomy and its associated sciences, this culminated in me producing STARGAZING with its wide-ranging subjects. It is a great joy for me to be able to share this knowledge with you.

I must also thank the scientists, science journalists, documentary makers, and the authors of my numerous science books that excited to me to learn since I was a child old enough to read 'How & Why' books. It always seemed that I could not get enough science books and magazines, or watch enough documentaries to satisfy my lust for knowledge of the big picture of things. I am indebted to the writers and producers of those exciting educational works because they allowed me to gather a wealth of knowledge to build an enormously valuable library of information.

I must also give a very special thanks to the scientists, engineers, and staff that have worked at NASA and the ESA for the truly incredible work they have done to make seemingly endless discoveries. They have unlocked so many mysteries, and in doing so they have exposed truly magical wonders in our solar system and many more in deep space. As a teenager, I was glued to the TV for each of the live feeds from the Apollo astronauts as they explored the Moon. I felt like I was there with them! And over the decades, to have seen what has been discovered by every mission to all the planets and many asteroids and comets, as well as the Sun, has been far beyond our collective imagination.

I cannot go without saying what an indelible impression the Hubble Space Telescope has had on me. What it has revealed has been nothing short of mind-blowing. And for it to show us such extraordinary beauty in all aspects of the universe, has been an incredible joy. To be aware of the astounding discoveries that NASA and the ESA have made, and are still constantly making, makes me feel very privileged and lucky to have been alive in this wonderful age of discovery.

All the people mentioned above have taught me parts of the underlying concepts behind what drives the most basic laws that control everything in the universe. And what an education this has been! Time and again, I had to let go of the common, simplistic beliefs we have been taught to see the more complex realities that control our world and the universe. I hope *STARGAZING* might convey some of these concepts to you so that you too can enjoy the wonder and joy that astronomy and science generally has given me. If I have done a good job with *STARGAZING*, you might pass this excitement on to others.

I am so thankful that I have grown up in '*The Lucky Country*' of Australia, with its wonderful natural environments, beautiful modern cities, leading-edge technology, and pollution-free environments and exceptionally clear skies and good weather. Having had the opportunity to have traveled the world several times, this has enabled me to enhance my education in science, and also to have met many leaders in science in different fields.

No one in the history of mankind has had the opportunity to be aware of all that we know today. This has occurred due to our rapid advancement in technology, which has led to all the amazing discoveries that just keep coming. We are indeed very lucky.

My greatest joy has been the opportunity to share the wonder of the universe with others and to have had the opportunity to learn from them. It is my hope that *STARGAZING* will allow you to do the same and also have much enjoyment in doing so.

Gregg D Thompson

Stargazers get magnificent views of the night sky when they go camping away from the light pollution of developed areas. Credit: Adrian Mascenon

ABOUT THE AUTHOR

It would be difficult to imagine anyone more interested in all fields of astronomy than Gregg Thompson. Right from his early childhood and throughout his life, he developed a strong interest in all forms of science, but astronomy excited him the most as it encompassed most of the other physical sciences, and it was the one where the most amazing discoveries were being made.

In his late teens, he built his first telescope. In his 20's, he used his love of innovative engineering to design and build a highly practical and versatile observing chair to mount his large binoculars. (See Volume 1 Chapter 8.) Similarly, his love of architecture led him to come up with a new design for amateur observatories. It was a split roll-off roof design that was much more versatile and practical than other types of observatories. (See Volume 1 Chapter 12.) And like most amateur astronomers, he coveted ever larger telescopes.

ASTRONOMICAL INTERESTS

Not content with merely looking around the sky, Gregg's ambition was to make scientifically-useful observations. Before the days of high resolution astrophotography, he applied his artistic talents to making accurate, detailed drawings of the planets, the Sun, and comets.

Gregg meticulously observed hundreds of galaxies searching for exploding stars (supernovas). He spent 12 years developing maps of star fields around 300 of the brightest galaxies. This allowed amateur astronomers to make visual discoveries of supernovas that randomly occur in galaxies. In 1989, Cambridge University Press UK published his work, which he co-authored with another amateur astronomer, James Bryan Jnr

from the University of Texas. James observed galaxies around the North Celestial Pole that Gregg could not see and assisted with the editing of the book. The book was entitled '*The Supernova Search Charts and Handbook*'. For this work, Gregg was awarded the Amateur Achievement Award from the Astronomical Society of the Pacific in the US (the world's largest body of professional and advanced amateur astronomers). He shared the award with his colleague Rev Robert Evans who discovered many supernovas visually with the help of Gregg's charts. His charts also received the 'Highly Recommended' status from the prestigious *Rolex Awards for Excellence in Science*. Gregg's supernova search charts enabled amateur astronomers to discover bright supernovas, as they occurred. Most supernovas were previously discovered on photographs long after they exploded, so it was not possible to observe how they developed. By discovering supernovas as they started to explode, professional astronomers could then make detailed observations of these events using large telescopes and even the Hubble Space Telescope, thereby enabling astrophysicists to better understand the physics underlying each type of supernova. Supernova discoveries were critical to enable professional astronomers to determine the size and age of the universe. This led to the discovery that the rate of the universe's expansion was unexpectedly increasing. (See Volume 4 Chapter 1, Part 5.)

Gregg also produced at the same scale as the supernova charts 1,000 photographs of small faint galaxies taken from the 48" Schmidt plates that amateurs could use to visually detect supernovas in galaxies more distant than those in his charts.

Gregg also observed and recorded the visual appearances of hundreds of nebulas and star clusters in the Milky Way, as well as those in the Magellanic Clouds. For this, he used his 310 mm and 460 mm telescopes.

In 1992, the editor of Australia's largest publishing house, Weldon Publishing, attended one of Gregg's themed, special-effects stargazing shows for guests at a major Australian country resort, Coolum Hyatt. Many people from outside the resort traveled long distances to attend his shows. The resort is located in a dark sky location. The editor was impressed with Gregg's knowledge of the night sky and the way he conveyed his enthusiasm for astronomy, as he showed his thrilled audience objects in space in his large telescopes. She liked the exciting way in which he presented facts about the universe complete with atmospheric special effects lighting and a very large, specially-constructed screen on which he projected video vision of flying past the planets and into objects in deep space. She thought his fibre-optic lit Star Trek-like control center and his spacey sound effects added a lot to the atmosphere of the evening's experience. In view of this, she asked him to write a popular book on observational astronomy. It was titled '*The Australian Guide to Stargazing*'. It was published in 1993. To his amazement, the book was widely lauded and made the top-seller list. With new additions and reprints it has sold consistently for 25 years.

Right: '**The Supernova Search Charts and Handbook**' provide a means for amateur astronomers to discover exploding stars in hundreds of distant galaxies.

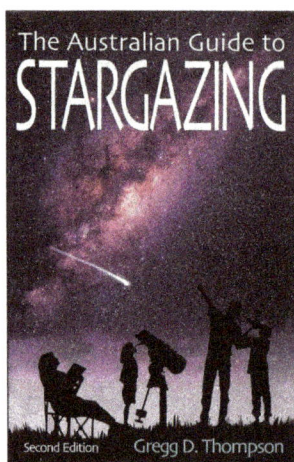

Left: '**The Australian Guide to Stargazing**' was a best seller.

Gregg's lifelong commitment to astronomy brought him into contact with many novice and advanced amateur astronomers, as well as professional astronomers, astrophysicists, and those in the media who promote astronomy. He organized, and participated in, many star parties at dark sky locations, as well as orchestrating tours of the night sky for members of the public at eco-resorts, and for guests on cruise ships on which he lectured. Gregg has been an invited guest speaker at astronomy conferences, club meetings, and business functions. He has taught educational courses in astronomy, and he has had many groups and schools visit his observatory. He has regularly been a guest on talk-back radio shows and TV programs where he would discuss a broad diversity of astronomical subjects and what was current phenomena to observe in the sky. He also spoke on how to control light pollution and he would explain what perceived UFO sightings actually were. Gregg's intellectual contribution to astronomy has been well-recognized.

BUSINESS PROJECTS RELATED TO ASTRONOMY

In business, Gregg is an accomplished entrepreneur. He has served as the managing director of a number of companies he has founded. One of Gregg's companies created promotions for national and multinational companies. Some of these have been based around astronomical themes or phenomena.

For the return of Halley's Comet in 1986, Caltex Oil (a division of Exxon) sponsored very successful educational charts and planispheres for this event. Gregg designed, produced, and successfully marketed these in Australia, New Zealand, the USA, Hong Kong, and Japan.

THE INFINITY EXPERIENTIAL IMMERSIVE MAZE

For another of his companies, Dreamtech Designs & Productions, Gregg designs, constructs, and operates multi-million-dollar themed leisure and edutainment attractions. Gregg's highly-acclaimed, experiential INFINITY attraction features a journey through a series of 20 multi-sensual, fun-filled, immersive art environments, which appear to recede to infinity. This 21st C funhouse features innovative special effects in immersive art environments. Its illusions push the boundaries of experiential themed attractions to the limit of technology. For instance, in the peaceful Star Chamber, patrons walk into heaven at the center of the galaxy where millions of stars extend for as far as the eye can see. Angelic choral voices make one feel like they are in heaven. In another environment, people laugh and dance on a waveform floor in the dark with only the light of colored starlight and their hands and feet glowing. In an endless disco-like environment, large spheres appear and disappear as they are bounce around in slow motion to ultra-groovy music. INFINITY's environments elicit a broad range of emotions. This has made it a successful attraction that is very popular for all ages.

THE SPACEWALKER EDUTAINMENT EXPERIENCE

In 2005, Gregg also created the *SpaceWalker* edu-tainment attraction, which allows visitors to discover the universe in many unique ways. This science-based attraction creates a high level of visitor interaction and participation. It utilizes leading edge special effects and highly themed environments that are suggestive of popular science fiction movies to give visitors many fun experiences, as they learn remarkable things about the cosmos.

As examples of his command of detail in the solar system section of the 'Space Walk Highway', Gregg created very realistic models of the planets, which appear to be floating in space as they rotate! Even special effects experts were not able to work out how he achieved this. He did not use holograms but real models. Moons orbit the Gas Giant planets with no apparent means of support. They cast their shadows on their planet's cloud tops, as they transit their planet's globe. Patrons can observe comets, asteroids, and space probes traveling through the solar system.

As Space Walkers journey through the spiral arms of our Milky Way galaxy, they see faithful, three-dimensional recreations of all types of deep sky objects. Gregg developed a means of constructing a three-dimensional globular star cluster with thousands of stars that appear to be suspended in space. Patrons pass by exo-planets orbiting other stars, as well as rapidly rotating neutron stars, and matter spiraling into a black holes. A supergiant star can be seen losing its outer atmosphere to a brilliant white, super dense, dwarf star in orbit around it.

After leaving our galaxy, Space Walkers travel into the depths of intergalactic space where they observe clusters of galaxies in every orientation that extend to the edge of the observable universe. Here, they can enter a rotating black hole where they travel through time. Or, they can take a return journey to the space station via a monster star gate filled with stars. Halfway through, Mission Control advises that aliens have invaded the *Stargate* as it drops into darkness. Patrons have to avoid being eaten by the monsters from the movie '*Alien*'.

In either case, visitors emerge safely at the entrance of the futuristic Restaurant at the End of the Universe to experience creative taste sensations from across the galaxy. Food and drinks are presented in very original ways where they light up and change color.

Many visitors to SpaceWalker commented on the extraordinary attention to detail and the imaginative concepts used. Images have been included to help readers appreciate the level of detail, the innovation, and the powerful means of communicating science to the public that Gregg has used. He has a talent for making learning about the universe exciting.

Gregg is a polymath due to his love for all the sciences – physics, chemistry, geology, meteorology, biology, neurology, psychology, sociology, logic, and of course astronomy. And because the universe includes life, astronomy indirectly includes the sciences that deal with life, us, and intelligence.

Over decades, Gregg has had considerable opportunity to discuss all manner of subjects related to astronomy with numerous people from all walks of life. This has given him a great appreciation for the sorts of subjects that interest amateur astronomers and general readers who have a broad interest in astronomy. He has included a wide diversity of subjects related to astronomy throughout *STARGAZING*, and especially in the last chapter.

When Gregg shows city dwellers the stars under a naturally dark sky, they are typically amazed at the vast number they can see. They are also surprised by how many meteors and satellites are visible. They are impressed with the easy visibility of the Milky Way and its delicate beauty. Most people today have never seen the Milky Way due to light pollution. It's only under dark skies that one gets to see the elusive Zodiacal Light and many faint meteors.

Gregg describes the effect that astronomy has had on his life by saying *"Once the doors to our imagination and curiosity are opened, we discover that the universe overflows with wonder, beauty, and mystery."*

Astronomy has given Gregg much enjoyment throughout his life, so it is his hope that he can repay what he has gained by sharing what he has learnt and experienced with *STARGAZING's* readers. They too will gain the sense of awe and wonder that he has experienced throughout his quest to discover the marvels of the universe.

Gregg thinks that life is going too fast and that in time, he will run out of years, so before that happens, he wanted to use all he had gained from his life in astronomy to write the ultimate book on stargazing for those interested in observing, and learning about, all aspects of the universe. As he has no need to profit from this project (that has taken him seven years to complete) he wanted to sell it at a low price, even though this would be unlikely to recover the costs involved in producing it, and then publishing it, let alone his time. He has done this so that it would be easily affordable to as many people as possible. He did not want to take this knowledge to the grave, so he hopes readers will agree that he has made the right decision.

Left: **Gregg's father bought him his first small telescope when he was 11. It was a very well used, terrestrial spotting telescope with a 60 mm (2.5") diameter lens and a magnification of 30 times. What he saw through it captivated his imagination so much that he became very enthusiastic about learning all he could about astronomy. He was also given the Larousse Encyclopedia of Astronomy which enthused his to want to see everything in that large book.**

Center: **By his early teens he had upgraded to a 60 mm Unitron refractor that permitted him to use magnifications up to 300 times. It also had very convenient slow-motion controls and a view finder.**

Right: **In his mid-teens, he worked through his school holidays to purchase a 114 mm (4") Royal reflector.**

Left: **In his late teens under the guidance of a beloved mentor, he learnt to grind and figure a 200 mm (8") f 6 mirror for a Newtonian reflector, which he mounted on a car axle.**

Right: **In his 20s, he used a 200 mm (8") f8 high resolution reflector telescope, which was made by his close friend and expert amateur telescope maker, Cliff Duncan who figured many mirrors to perfection. Cliff constructed his own foundry to make specialized parts for the telescopes that he generously designed and built for Gregg and other amateur astronomers at no cost - just for his love of the art of telescope making.**

Left: **In Gregg's 30s, he acquired a German mounted, high resolution 310 mm (12.5") Newtonian reflector.**

Center: **A decade later he acquired a 460 mm (18") high resolution Dobsonian reflector.**

Right: **Throughout those years, he used his very practical binocular chair for observing with his large 20 x 80 binoculars.**

INFINITY EXPERIENTIAL ILLUSION MAZE

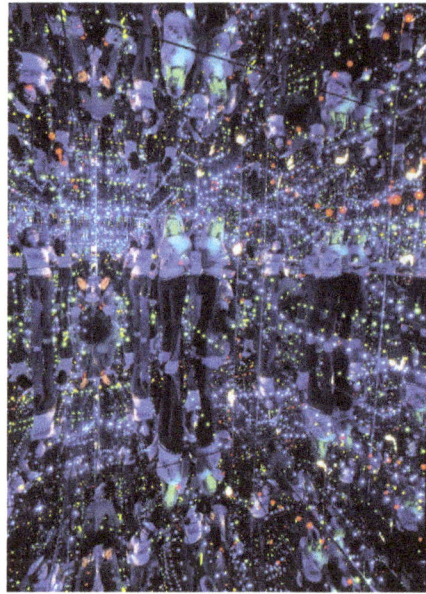
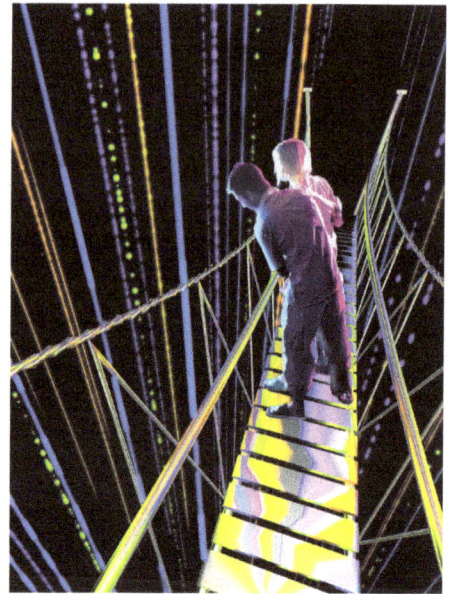

Left: **In Infinity's 'Infinite Kaleidoscope', people dance to music while immersed in a continuously changing kaleidoscope of colorful images that extend for 20 stories in every direction.**

Center: **Infinity's 'Electron Maze' simulates being inside the 'mind' of a gigantic computer where electrons are racing along circuits that never end in every direction. Each circuit has its own color and sound field. Patrons have to find their way out of this seemingly endless maze to get to their next experience.**

Right: **Patrons dare to cross Infinity's 'Light Canyon's' suspension bridge, which extends over a bottomless chasm.**

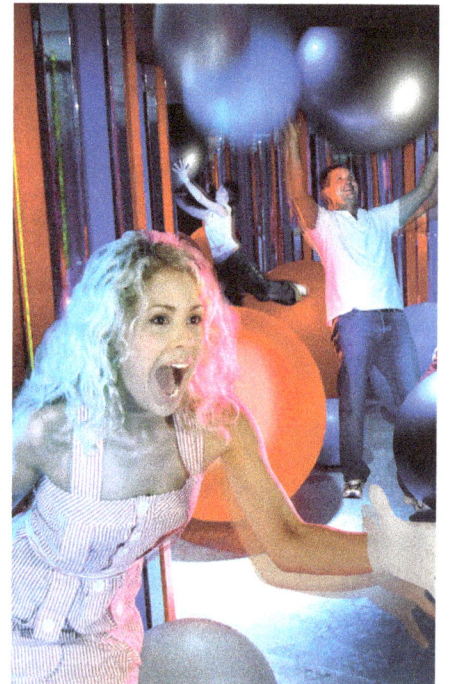

An infinite computer-like maze   **An infinite light chamber**   **An infinite disco of disappearing**

SPACEWALKER EDUTAINMENT ATTRACTION

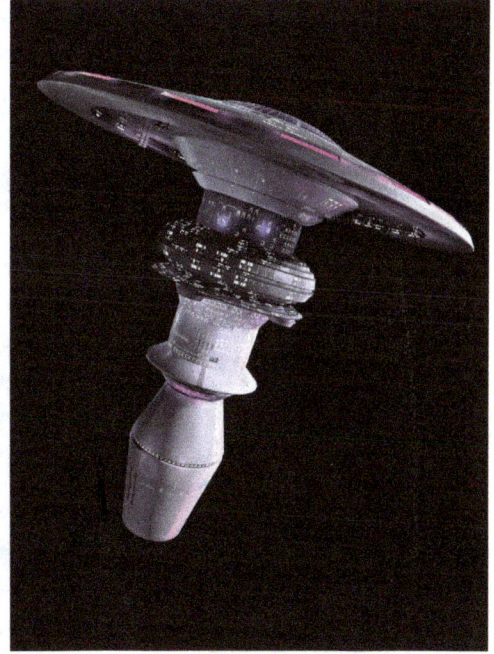

At Earth Station, visitors experience a simulation of being dematerialized and teleported to the giant Space Station Zeta out in the galaxy. Here, advanced aliens enlighten humans about amazing aspects of the universe.

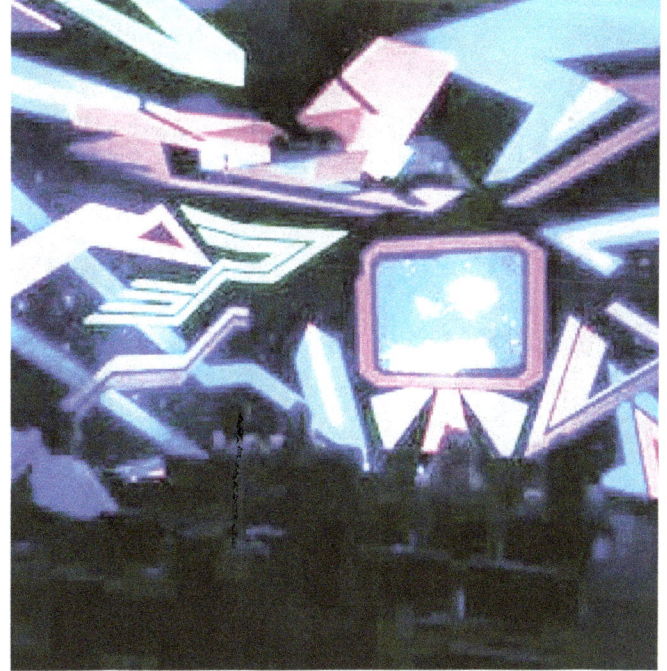

Left: **The Arrivals Centre** *at* SpaceWalker's Star Station Zeta

Right: This is Pod 1 where audiences are taken on a virtual reality flight over Venus and Mars's extraordinary landscapes.

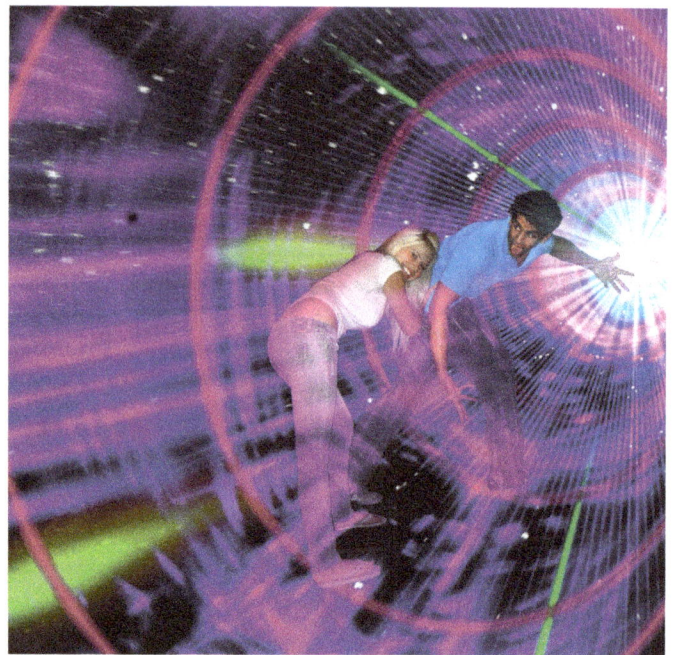

Left: **In one of SpaceWalker's Discovery Pods, humans are suspended in a 'mind transfer' unit that recreates a pleasant evening on Earth. An alien spacecraft descends from the stars to take Earthlings on a mind-bending, faster-than-light journey to the edge of the observable universe so they can learn about the structure of the cosmos.**

Right: **At the end of the universe, visitors travel through the spinning accretion disc of a black hole to go inside a space warp that transports them back to the space station's Restaurant at the End of the Universe.**

Left: **At the outer edge of the space station, spacewalkers can pass through an airlock into a rotating Stellar Vortex Wormhole (right).**

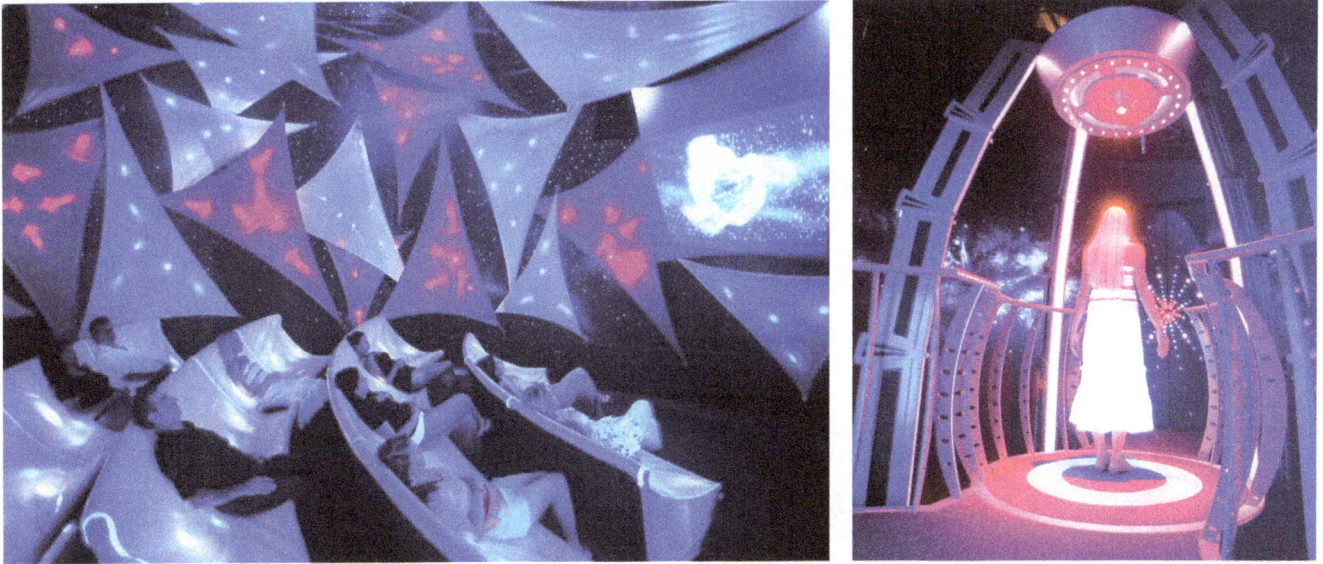

Left: **Here we see Discovery Pod 2, which takes spacewalkers on journeys to strange but real planets.**

Right: **At edge of the space station there is a Stargate where patrons take a leap of faith to jump into to experience micro gravity amongst a sea of stars in every direction.**

The Space Walk section realistically replicates a superb diversity of all types of objects in the universe. Travel along a 'highway through the universe' that goes through our solar system, then into the Milky Way galaxy, and finally into intergalactic space. Gregg created highly atmospheric sound fields to give each section a very 'spacey' feel to it. Along this highway are 'space-time viewers' in which visitors can watch exciting 1 to 2 minute videos of the most impressive features of each type of object in the universe. Gregg is justifiably proud of the quality of the three-dimensional models that he and his artisans created under his guidance. Each object was meticulously crafted to realistically replicate their real-world counterpart.

This extraordinarily accurate 3D model of the Moon took a master sculptor six months to sculpt all features in relief. Each of the Moon's numerous features had to be located exactly to perfectly replicate the detail seen in small telescopes. As one walks along the Space Walk highway, the Moon can be viewed from New Moon through to Full Moon, and due to its slow rotation, the far side of the Moon is also visible.

Left: **Jupiter and Saturn were very difficult to create because they had to be perfect oblate spheroids with their moons orbiting them at different rates. Jupiter's Galilean moons had to pass in front of the planet with no apparent means of support. Each moon cast its shadow on Jupiter's cloud tops - just as they do in the real universe.**

Right: **Saturn was especially difficult to light due to the gap between the rings and the planet's globe. The rings had to cast their shadow on the globe while the globe cast its shadow on the back of the rings, yet not spill any (sun)light. Like the real planet, the rings rotated, and Saturn's family of moons orbited the planet at different rates.**

AN INTRODUCTION TO THE STARGAZING VOLUMES

When talking to amateur astronomers from all over the world, over the decades I often heard calls for a comprehensive book that covers all aspects of stargazing. I believe the 4 Volumes of *STARGAZING* satisfy that need because they incorporate the following:

- All the **practical things you need to know to stargaze** for the joy of it as well as doing **serious observational astronomy**,

- Learning the basics about **how the atmosphere works** to maximize one's opportunity of observing under the best conditions,

- **Understanding how your eyes work to maximize our vision** to see very faint objects and to detect maximum detail, as well as knowing **how to test your eyesight**,

- **How to prepare for observing sessions**,

- **The advantages of observing under a dark sky**,

- **Which objects are best for viewing with the naked eye, binoculars, and telescopes**,

- Tips on **buying and using telescopes**, and other astronomical equipment,

- The types of **observatory designs** that are the most practical,

- **How to minimize light pollution**,

- **Finding one's way around the sky and understanding astronomical terms**,

- **Understanding the nature and the significance of what you are observing**,

- How to get started with **astrophotography**,

- How to go about **observing the Moon, the Sun, the planets, asteroids, and comets**,

- **Observing the best examples of all types of deep sky objects** (star clusters, nebulas and galaxies) in both the northern and southern hemispheres,

- Observing all types of **astronomical events** such as **eclipses** of the Sun and the Moon, **meteor showers**, **satellites**, **auroras**, and the **Zodical light**,

- Descriptions of the **world's major observatories** for stargazers to visit,

- **Drawings of astronomical objects** to illustrate what objects look like visually, and **how to make your own quality drawings**,

- **Excellent information graphics** for concepts, types of objects, and astronomical events,

- The **best and latest images** taken by large **professional observatories, space telescopes and space probes**, as well as those taken by **the most advanced amateur astronomers**. Impressive 3D images are included,

- **The best space art** by the world's leading CGI artists,

- The **latest discoveries** across each field of astronomy and related sciences with **easy-to-understand explanations**,

- **Personal anecdotes** about unusual observations and events,

- How to have exciting **'Wow!' experiences** when observing,

- **Observing challenges** that stargazers like to test their observational abilities,

- **A big picture view of the size, structure, and the evolution of the universe**,

- The likelihood of **life and intelligence** in the universe,

- **The latest concepts and discoveries about cosmology,**

- **The most remarkable aspects of the history of astronomy** so that readers can appreciate how much we owe to our greatest, intuitive visionaries,

- **Answers to the BIG questions** that people want to know that astronomers, physicists and cosmologists are discovering.

Because *STARGAZING* is so all-inclusive, it took seven years to write it and edit it many times over. Exhaustive research was done to ensure it was up-to-date with the latest discoveries. Creating many of the illustrations as well as finding and collecting all the images were also massive undertakings. Much time was consumed in managing numerous communications, preparing the layout for each chapter and creating the website. When all this was done, organizing the distribution of the book and its marketing were also major tasks.

Because the book is a large body of work that has a high content of superb, large color images, it would be prohibitively expensive to publish it as a traditional printed book. This would severely limit its potential readership. In view of this, I decided to make it an ebook to make the cost minimal so it would be appealing to a broad audience. This would allow me to share all that I have gained from astronomy with as many readers as possible.

STARGAZING'S EXTRAORDINARY IMAGERY

The book's page size was originally designed as a 260 mm x 300 mm (10" x 12") printed coffee table book, but as the book grew in size, it had to become an e-book. This allowed very large images to be included so they can be zoomed into to see remarkable detail.

Thanks to recent advances in digital photography and computer processing, amateur astronomers are now capturing excellent, high-resolution images, even of very faint objects. Because of this,

STARGAZING features the best **astrophotography taken by the world's most advanced amateur astrophotographers** from many countries. These images have a depth of color and detail that was unimaginable even a decade ago.

There are also stunning images taken by **large, professional, ground-based observatories, space telescopes, and space probes**. The very high-resolution images taken by the Hubble Space Telescope (HST) and the European Southern Observatory (ESO) are simply stunning.

The book's imagery also includes reasonably scientifically accurate, **space art** painted by a number of the world's leading space artists. Their illustrations allow readers to visualize what planetary landscapes, exo-planets, bizarre stars, and other solar systems might look like.

STARGAZING also features **drawings of astronomical objects** made at the telescope in both black and white, and in color. These drawings illustrate what objects appear like visually so that novices do not expect to see Hubble Space Telescope images in their telescopes.

STARGAZING's images convey far more information than written words could ever do, and they demonstrate the extraordinary beauty that abounds across the universe.

WHO IS STARGAZING WRITTEN FOR?

STARGAZING's volumes are written for anyone interested in astronomy:

- **novice stargazers** who want to know how to go about observing and what exciting things they can see,

- **experienced amateurs** who want to know more about the physics of the objects and the phenomena they are observe,

- **advanced amateurs** who want to broaden their general knowledge into areas that they have not delved into,

- **professional astronomers and many scientists** who are so consumed in their field of expertise that they find it hard to keep up with discoveries beyond their specialty. The latest concise discoveries in STARGAZING are appealing to all readers,

- **astrophotographers** who love to see the most extraordinary imagery in every field of astrophotography,

- **students and science teachers** who find STARGAZING especially good for learning about all aspects of astronomy and its associated sciences,

- **general readers** who love learning about the cutting edge of astronomy and cosmology,

- **futurists** who want to visualize what the near and far future will be like, and the role that technology, intelligence, and consciousness may play in the evolution of the universe.

STARGAZING'S MAGIC MIX OF SUBJECTS

STARGAZING contains **exciting explanations of many subjects**, together with excellent **info-graphics** about **the characteristics of all types of astronomical bodies and the diversity of phenomena in the universe.** The text describes how such things work and evolve in **easy-to-understand language**. As well, it has numerous **thought-provoking concepts** supported by **helpful analogies** that make them easier to understand. The book is interspersed with surprising historical facts, and amusing personal anecdotes. Where applicable, it also includes references to **science fiction movies and documentaries** that are related to the subjects being discussed.

I am very lucky that astronomy has given me more than my fair share of 'Wow!' experiences, particularly when I have been observing at very atmospheric locations and when witnessing awe-inspiring astronomical phenomena. I have shared these experiences throughout the book, so that others might delight in having similar experiences.

Most readers should find *STARGAZING's* magic mix of diverse subjects combined with its exciting style of presentation, to be a thrilling and pleasurable read.

WHAT YOU SHOULD KNOW ABOUT THE STARGAZING TEXT:

- It requires **no specialist knowledge of science or mathematics,**

- **Names, technical terms, and important words are shown in bold** so they can be located quickly when scanning the text.

- Many sections have **summaries of a section shown in light blue,**

- The author's **anecdotes are shown in navy blue,**

- **Challenging questions** for readers to contemplate are shown in violet. When there is an observational challenge, this is indicated by the icon **C** . These challenges are given because many amateur astronomers like to test their observational skills and their instruments,

- Throughout the text, are **information 'boxes'.** These have a blue-gray background. These are included to provide additional interesting information that is indirectly related to the topic of discussion,

- **Most quantitative information is presented as rounded off numbers** because many measures and quantities in astronomy are estimates. Rounding off numbers makes them easier to remember,

- **Measurements** are given in Metric units with Imperial units following in brackets. Kilometers are abbreviated to **km**, meters to **m**, millimeters to **mm**, miles to **mi**. Feet are denoted by ', and inches by ". Minutes of arc are denoted by ' and seconds of arc as ". Kilograms are abbreviated to kg, and pounds to lb. Temperature is abbreviated to C for Centigrade, F for Fahrenheit, and **K** for Kelvin,

- **Major well-known professional telescopes, observatories and Space Agencies** are abbreviated as follows: Hubble Space Telescope as **HST**, National Aeronautics and Space Administration as **NASA**, the European Space Agency as **ESA**, the Japan Aerospace Exploration Agency as **JAXA**, the European Southern Observatory as **ESO**, and the National Optical Astronomy Observatory as **NOAO**.

- **Star charts** and **Celestial coordinates** of deep sky objects are not provided because they are seldom required nowadays due to most modern amateur telescopes using computer-aided object finding technology (CAT systems). If coordinates are required for 'fixed' objects, such as stars, nebulas, and galaxies, they can be found using night sky Apps or a smartphone. They are easy to find by going on the Internet and typing in the object's name. Coordinates for moving objects such as comets and asteroids can be found online by searching for the body's name and appending the word 'coordinates'.

In the vein of Carl Sagan's extraordinarily popular 'Cosmos' series, *STARGAZING* includes some **thought-provoking speculation** about many subjects. The speculation in *STARGAZING* is founded on recent scientific discoveries and through making logical deductions.

The rate at which major discoveries in all sciences are racing ahead, is forcing us to rethink many things that that we thought were solid facts and commonsense. In some cases, we are now having to question what is reality. As we learn more with an open mind, we have to be prepared to let go of old, simplistic beliefs and consider the evidence that exists for some very complex aspects of nature. Some of the conclusions that we are confronted with seem impossible at first, but once understood, they are intellectually very stimulating and they give us a new view of nature. The evidence that supports them makes them hard to refute. This is particularly the case for many subjects in Volume 4.

Since I was a child, I have had a lifelong love for astronomy due to the awe that it has inspired in me. Over the years that *STARGAZING* has taken me to compile it, I have learned a great deal about many mind-expanding discoveries that are at the leading edge of astronomy, and I have become aware of amazing new technologies that will dramatically change our future. As well, I have discovered truly incredible aspects about the subjects that are involved in the evolution of our universe. Putting *STARGAZING* together has been a very exciting and stimulating cerebral journey for me. As you read through it, you too will embark on this amazing journey. Not only will it open your mind to incredible concepts, you will see unbridled beauty. You will also time-travel back to the very beginning of the universe's creation and into its distant future!

I hope you will enjoy the 'ride' you are about to take and the knowledge that you will gain as much as I have.

Gregg D Thompson

OUR STRANGE AND COMPLEX COSMOS

In this chapter, we surf the waves of cosmic concepts.

CHAPTER 1

COSMOLOGICAL CONCEPTS & ENIGMAS

This chapter looks at a special branch of astronomy known as cosmology. It deals with the astonishing origin and evolution of the universe, its enormous scale, and its perplexing structure. We discuss what its purpose might be? Will it age and cease to exist? Does it go on forever? Is its geometry of space flat or curved and how many dimensions might it have?

To appreciate what underlies cosmology, in Chapter 2 we investigate the **four known forms of energy, as well as the nature of visible matter**. In Chapter 6, we look at whether **the recently discovered dark matter and dark energy really exist.**

In Chapter 3 we examine **how the universe is *thought* to have come into existence as an infinitely hot, dense point that was trillions of times smaller than a proton and how it expanded extremely fast to grow** into the enormous size it is today. Cosmologists are confounded about how our universe appears to have been born out of nothing! In Chapter 11, we discuss two likely ways it may eventually die.

In Chapter 12, **we ponder the very important role that the opposing forces of order and disorder play in the evolution of everything in the universe. We'll discover what role chaos plays. It was once thought to be disorder but is now known to be an unsuspected higher level of complex order. It's a simple formula that drives the evolution of everything in the universe.**

Chapter 14 explains a startling aspect about the universe – one that so far defies any explanation. Cosmologists have found that **there are some 30 physical parameters in our universe that are *incredibly* finely-tuned to a level exceedingly beyond the possibility of chance!** Without this miracle of fine-tuning, our universe would never have evolved, let alone produce life and intelligence. How this came about without it being designed is incomprehensible.

Particles of matter are thought to be incredibly small packets of energy known as quarks that interact with one another and with the four force fields of electromagnetism, the weak and strong nuclear forces and gravity. The strong nuclear force glues quarks together so strongly that enormous amounts of energy are needed to separate them even a little. The further apart they are forced, the stronger the strong nuclear force becomes. If they are separated, this energy creates two new quarks!

In looking at the structure of the universe, we realize that the speed of light determines much about our universe. All that we know about our cosmos is contained in an enormous 'bubble' that we call our *'observable universe'*. It exists in the enormously larger **Greater Universe**, but because this lies beyond the horizon of our observable universe, there is no way of discovering anything about it because no information from it can reach us due to space across the Greater Universe expanding faster than the speed of light. (See Chapter 4). (Space can expand faster than light but matter cannot).

In Chapter 13, you'll marvel at the recently discovered and completely unsuspected **froth-like filamentary structure of galaxy clusters throughout the universe.** Astronomers have discovered that **the poles of supermassive black holes** that lie at the centers of all galaxies, are aligned perpendicular to the filaments in which they lie! What force could possibly cause this to occur across the whole universe? These two discoveries are extraordinary! There is a surprising possible explanation for the first, but not the second.

Part 7 ponders whether it is likely that there could be other **parallel universes, or a multiverse**.

Cosmology raises many mystifying questions. It delves into the most thought-provoking concepts that are the most intriguing that the human mind has ever encountered. The more science discovers about the cosmos, the more complex, bizarre, and seemingly counter- intuitive it appears to be. We need to learn much, much more to be able to understand its complexity at every level.

Astronomers and cosmologists are hard at work developing new theories to explain a never-ending stream of new discoveries. Unexpected new observations are continually presenting new challenges to our scientific beliefs and forcing us to think more laterally. Solving these riddles often requires the development of totally new concepts and the development of an array of new technologies.

CHAPTER 2

HOW THE UNIVERSE EVOLVED SIMPLIFIED

To understand how the universe evolved, it is important to have a basic understanding of the most common particles of matter and the four forces of nature that act upon them. In this section, there is no in-depth physics or mathematics, so you should have no trouble in comprehending most concepts. If a section seems too much for your interest level, then you can gain a basic understanding by looking at the images and diagrams and reading their captions.

To comprehend how the universe works, it is necessary to realize that **gravity is the basic organizing force in the cosmos.** It is crucial in creating the structure of everything in the universe. Gravity is an incredibly feeble force on small scales. It is 10 billion, billion, billion, billion times weaker than the other three forces (electromagnetism and the strong and weak nuclear forces). Gravity acts over large scales on everything in the universe. It is what causes mass to collect together to form stars, planets, galaxies and huge superclusters of galaxies. But gravity is not actually a force! It occurs as a result of mass (matter) curving space and slowing time, just as Einstein predicted and later this was proven to be true. (See Volume 2 Chapter 6, page 198.) As it curves space, it forms gravity 'hollows' which causes matter to fall into these depressions. The greater the gravity, the deeper the depression into which matter falls.

In simple terms, the early stages of the expansion of the universe occurred something like this. In an extremely small fraction of a second after the universe started to expand, known as the **Big Bang,** it is thought that the four forces of energy were unified as one force. At this time the universe was almost infinitely hot. As it rapidly expanded, its temperature decreased. In doing so, the four forces decoupled one after the other. This allowed the most basic forms of matter - quarks and their antiparticles - to 'freeze out' of the strong and weak nuclear forces. To visualize this, think about how a bottle of liquid soft drink near freezing point freezes instantly when the pressure is released when the cap is taken off.

At this point in time, it is postulated that the universe must have experienced an incredibly fast expansion (called inflation) to spread out the matter exceptionally evenly. This is thought to be the reason why had to happen so that the universe was not lumpy - which it isn't. This extremely rapid expansion expands 100 trillion, trillion, trillion times (- that's 10^{-26} times, or 10 with 26 zeros after it) in a minute fraction of a second (only 10^{-33} to 10^{-32} of a second)! In that time, it went from being as small as a proton to the size of a grain of sand or maybe a grapefruit in an instant!! This might not sound like much, but it is an incredible rate of expansion in almost no time at all! (It should be stated that this means of spreading matter evenly across the universe is speculation at this stage.)

As the theory goes, the incomprehensible rate of inflation caused rapid cooling which caused quarks and a multitude of subatomic particles to 'freeze out' of the force fields. With the universe's expansion continuing,

A BRIEF HISTORY OF THE EVOLUTION OF MATTER

The following is a short summary of how the universe evolved into increasingly more complex forms to create life, and ultimately, intelligence and conscience.

At the birth of the universe, it was so hot that only energy existed. As space expanded to become much larger, this caused the universe to cool. It is thought that there was no matter until some of the energy 'froze out' into extremely tiny particles known as **quarks.** As it cooled more, quarks combined to form around a thousand small, **sub-atomic particles** - the main ones being protons, neutrons and electrons. As the temperature continued to drop, these particles combined to form **atoms.** They formed chemical bonds with one another to become **molecules** of gas and dust. Gravity then drew matter together to form **stars and galaxies** which were then gravitationally drawn together to form clusters of galaxies.

On planets with the right conditions, complex arrangements of molecules formed complex organic compounds **that could replicate** like crystals do,

it cooled rapidly to a point where the particles such as protons, neutrons, and electrons could come together to form atoms. Gravity caused atoms to collect together to form stars and solar systems were then able to form. A detailed account of this evolution can be found on page 1150. In the near future, quantum computers will allow physicists to understand how the early evolution of the universe came about in much more detail. It will be much more complex than our present assumptions.

Matter evolved to create everything we know of in the universe, so let's now look at how it managed to become ever more complex. The following diagrams provide a simple view of the four main states of matter – solids, liquids, gases, and plasmas. These states are determined by temperature and pressure. Understanding this will help you visualize phase changes, as occurs when steam changes to water, and then to ice as the temperature drops and/or the pressure decreases.

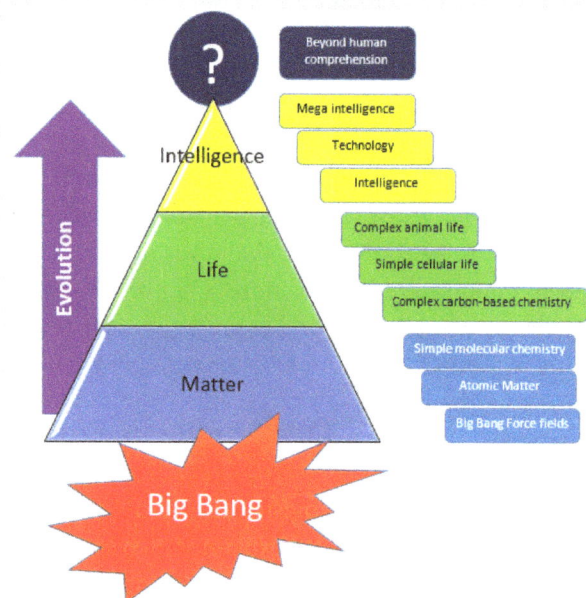

The evolution of the universe - as far as humans can determine to date.

but in far more complex ways. These chemical structures formed **amino acids, RNA, and later DNA.** These molecules then evolved into much more complex viruses, which were not quite what we call life. Eventually much more complex **single cell organisms** formed the first life forms. They became the basis of more **complex multi- cellular life forms.**

4

These evolved into much more complex forms of life that developed a **broad diversity of ever-improving senses, as well as means of movement, digestion and defenses.** The most advanced life forms evolved **levels of intelligence and memory** that became ever more complex. Ultimately, the most advanced **intelligences developed high technology.** This caused the evolution of organic life forms to speed up dramatically. They quickly learned how to manage genes to make themselves much more advanced and to also build other life forms through genetic engineering. They also discovered how to expand their minds beyond their brain by inventing computers, which quickly evolved into a **non-biological super-intelligences** with unbounded potential. This is the level of evolution that humans are now experiencing.

If life is common throughout the universe, as it appears to be, then there is little doubt that there would be innumerable intelligent life forms that have evolved over billions of years before life on Earth even started to evolve. By now, they would have evolved immeasurably far beyond our state of consciousness.

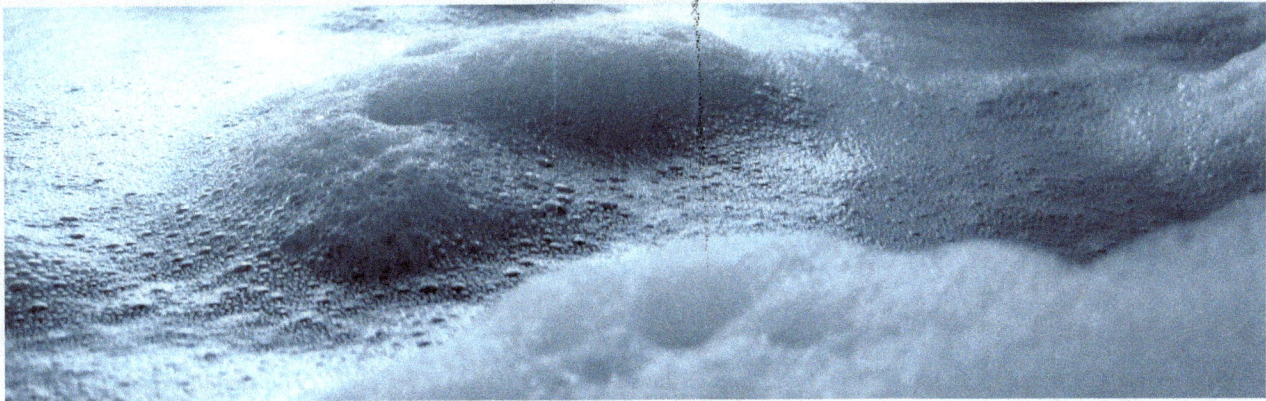

At the smallest sub-atomic scale, background energy fields, even in a vacuum, are thought to be like a landscape of energy 'bubbles' spread throughout all of space. Credit: Gregg Thompson

THE BASIC COMPONENTS OF MATTER

The following sequence illustrates how matter evolved. Because it is unknown what particles of matter really look like, the following diagrams are purely conceptual to help readers visualize how matter may have formed. (All diagrams by Gregg Thompson and Nicole Brooke.)

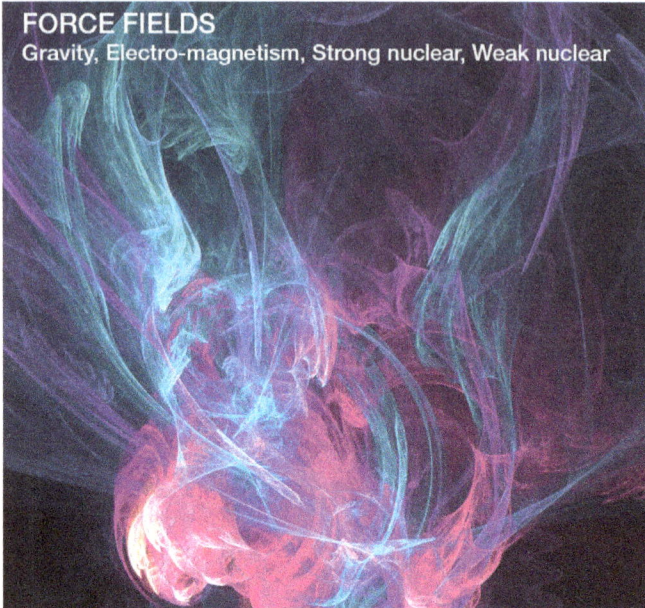

FORCE FIELDS
Gravity, Electro-magnetism, Strong nuclear, Weak nuclear

After the Big Bang occurred, four force fields 'froze' out of the background unified field and started to cool as the universe expanded.

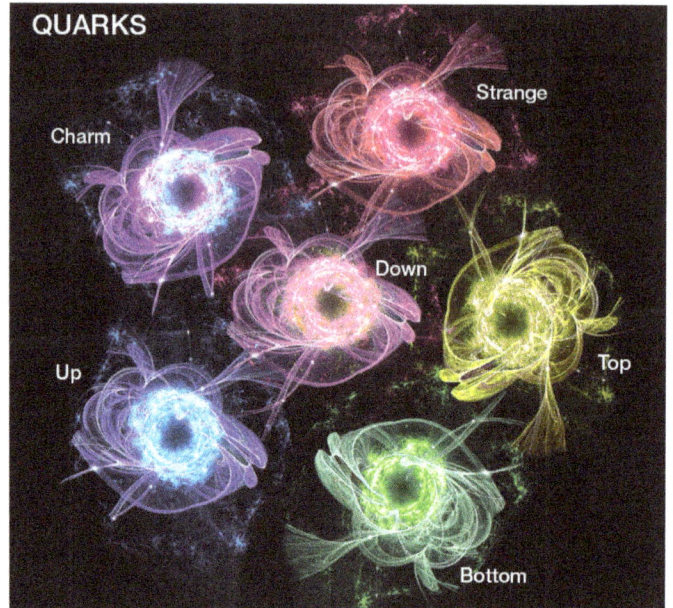

QUARKS

Charm, Strange, Down, Up, Top, Bottom

Three of the force fields cooled, allowing the smallest particles of matter - six quarks - up, down, top, bottom, charm, and strange - to 'freeze' out.

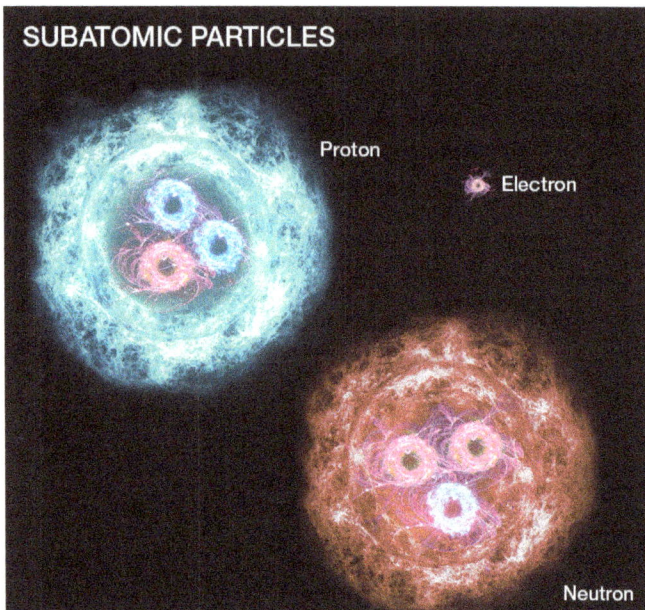

SUBATOMIC PARTICLES

Proton, Electron, Neutron

Subatomic Particles - quarks combined to form nucleons, protons and neutrons. Protons (the turquoise outer haze) have two Up quarks (light blue) and one Down quark (light pink). Neutrons are the red outer haze. They have one Up quark and two Down quarks. Electrons remained free at this stage.

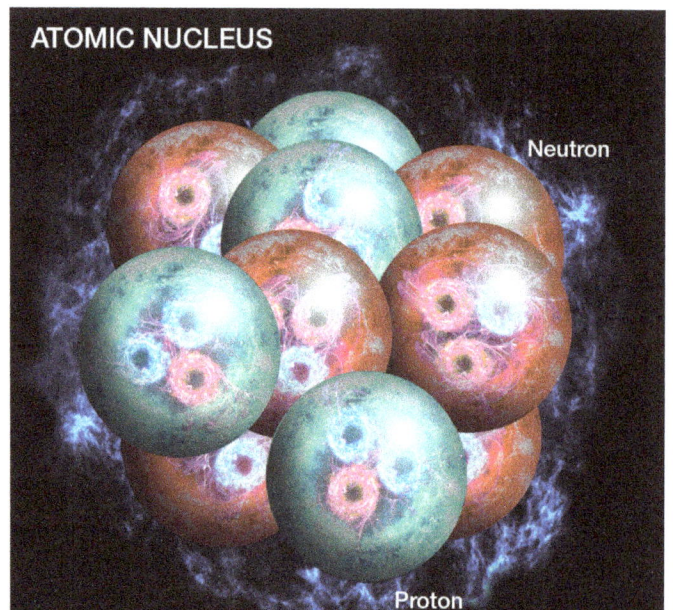

ATOMIC NUCLEUS

Neutron, Proton

Protons and neutrons combined to become the nuclei of atoms. When the temperature dropped enough, they were able to capture electrons to form atoms. Atoms are typically represented as spheres for simplicity, but they are probably not like this in reality.

6

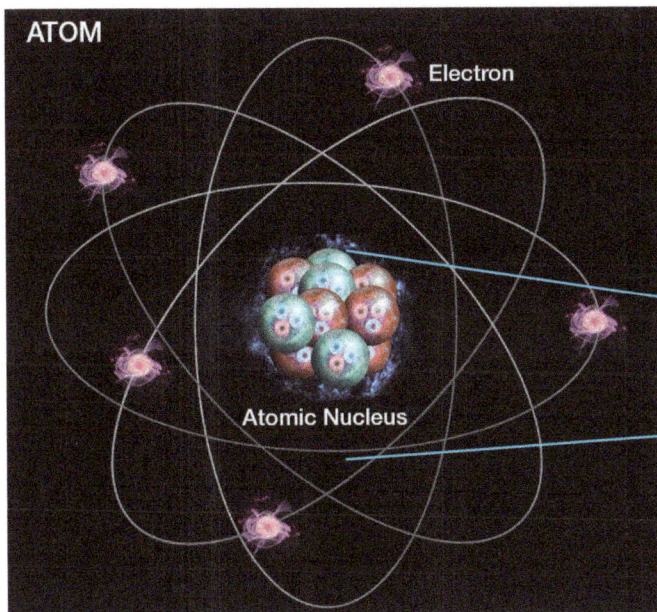

ATOM

Electron

Atomic Nucleus

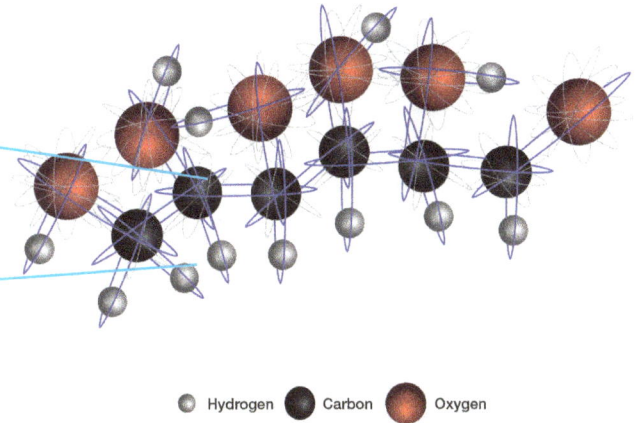

GLUCOSE MOLECULE
(SUGAR)

○ Hydrogen ● Carbon ● Oxygen

Electrons 'orbited' atomic nuclei to form atoms, but in strange ways we do not understand, and not like they are represented here. Electrons could whiz around more than one atomic nucleus to bind atoms together to form molecules. In reality, electrons do not 'orbit' a nucleus like a satellite does around the Earth. They are more like very fuzzy clouds that can be in many places at once. Atoms are so small that 10 million of them would be required to sit end to end to span the full stop here!

Atoms of carbon (black), hydrogen (gray), and oxygen (red) have combined here to form this molecule of glucose (sugar). Each atom shares its outer electrons (blue ellipses) to hold the molecule together. The more energy atoms have, the more they vibrate. If they have too much energy, the molecule 'shakes' apart into its basic atoms.

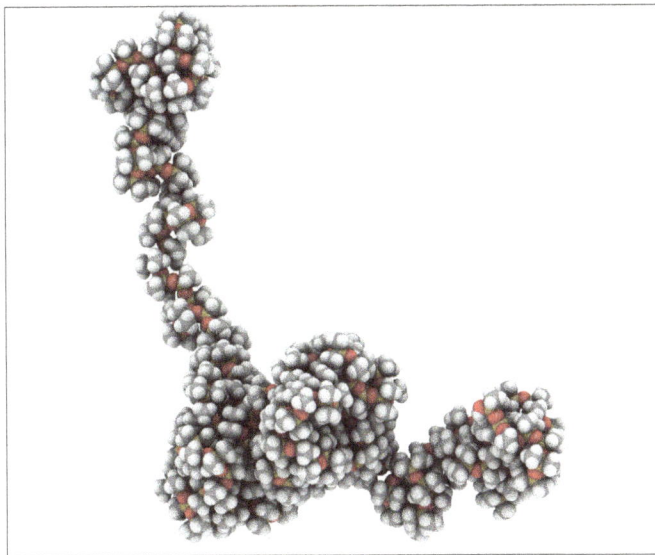

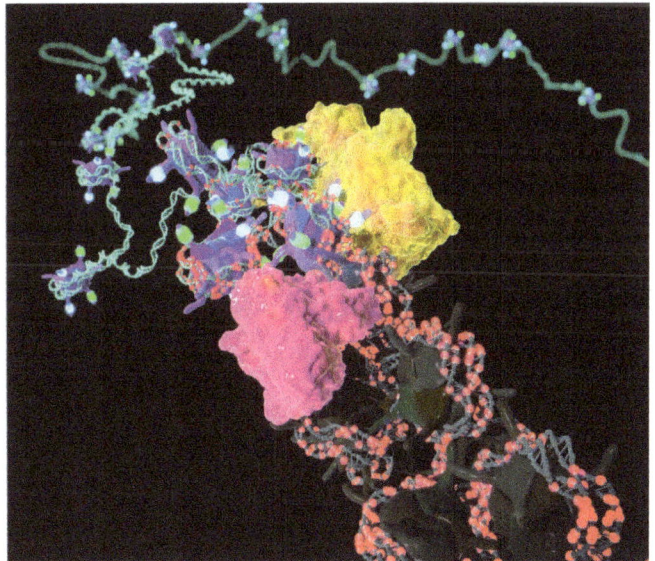

Numerous molecules can combine with carbon to form giant chains of organic molecules knowns as polymers. These become the foundations for constructing life.

Nature uses long carbon polymer chains to build very complex machinery inside cells to allow them to build and dismantle the molecular structures they need to have senses to find food, and to process it to get energy, as well as eject waste, to defend themselves, and to replicate. Credit: Ricken Research

Standard Model of Elementary Particles

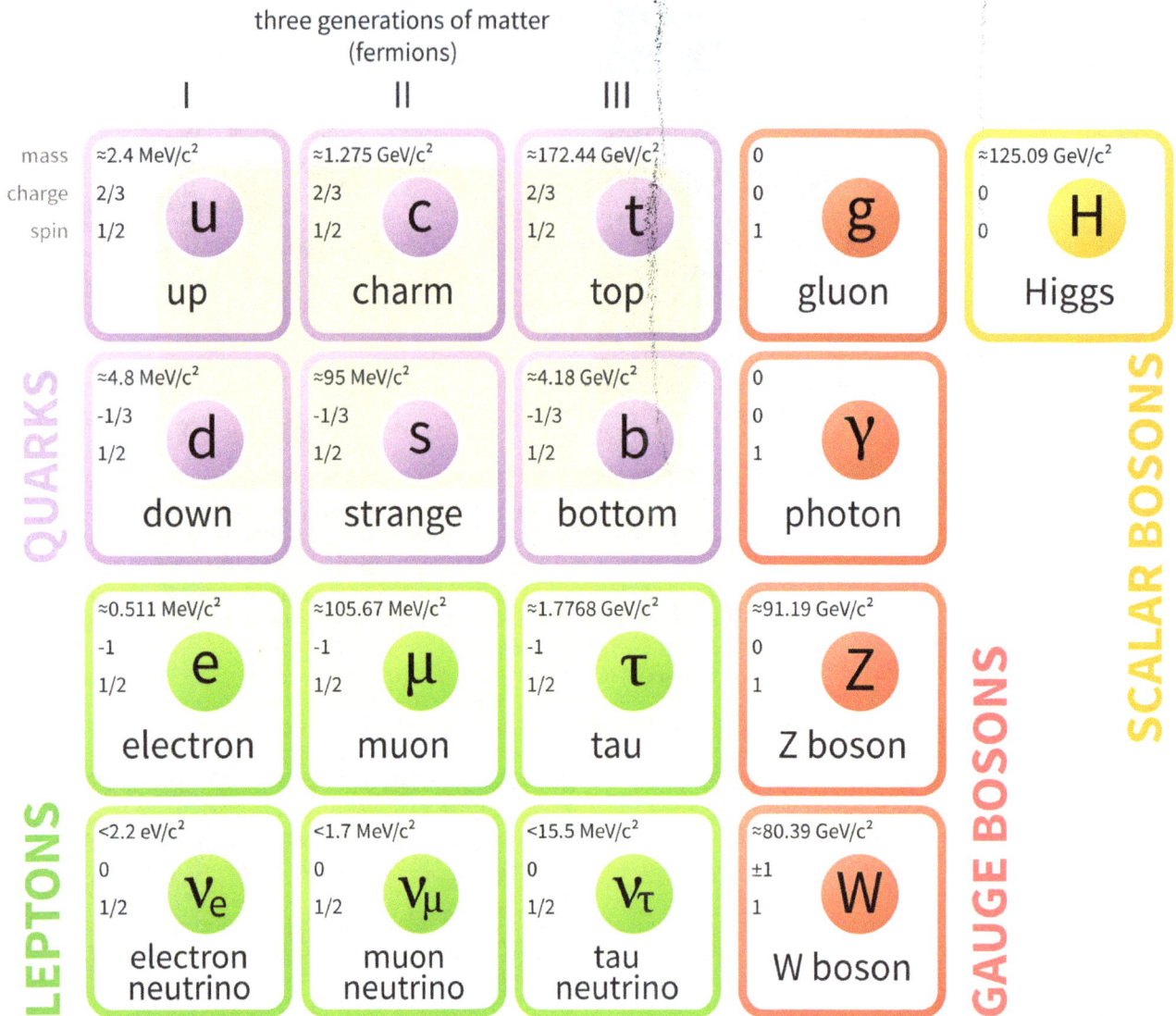

three generations of matter
(fermions)

I   II   III

QUARKS

mass $\approx$2.4 MeV/c^2	$\approx$1.275 GeV/c^2	$\approx$172.44 GeV/c^2	0	$\approx$125.09 GeV/c^2
charge 2/3	2/3	2/3	0	0
spin 1/2 **u**	1/2 **c**	1/2 **t**	1 **g**	0 **H**
up	charm	top	gluon	Higgs

$\approx$4.8 MeV/c^2	$\approx$95 MeV/c^2	$\approx$4.18 GeV/c^2	0	
-1/3	-1/3	-1/3	0	
1/2 **d**	1/2 **s**	1/2 **b**	1 **γ**	
down	strange	bottom	photon	

LEPTONS

$\approx$0.511 MeV/c^2	$\approx$105.67 MeV/c^2	$\approx$1.7768 GeV/c^2	$\approx$91.19 GeV/c^2	
-1	-1	-1	0	
1/2 **e**	1/2 **μ**	1/2 **τ**	1 **Z**	
electron	muon	tau	Z boson	

<2.2 eV/c^2	<1.7 MeV/c^2	<15.5 MeV/c^2	$\approx$80.39 GeV/c^2	
0	0	0	$\pm$1	
1/2 **ν_e**	1/2 **ν_μ**	1/2 **ν_τ**	1 **W**	
electron neutrino	muon neutrino	tau neutrino	W boson	

GAUGE BOSONS

SCALAR BOSONS

This chart shows the energy levels, the spin, and the charge (in small black type) for each type of known particle. For each quark and lepton, there is an antiparticle.

HIGH ENERGY QUARKS	MEDIUM QUARKS	LOW ENERGY QUARKS
Top and Bottom	**Charm and Strange**	**Up and Down**

Quarks can't be isolated. They must always exist in pairs. The more they are pulled part the greater their energy grows because they absorb the energy used to pull them apart. If they are pulled a long way apart, new quarks form in the space between them. There are three versions of quarks and leptons with each having more massive versions of the same particle for reasons we do not understand. The heavy versions only exist in high energy events like the Big Bang.

Neutrinos are almost massless particles that move at almost the speed of light. They are formed in the cores of stars in huge quantities. They do not interact with matter. It would take about a light year of lead to stop them. Sixty-five billion pass through our body every second!

Because a photon of light travels at the speed of light, it does not experience time, so to a photon, the moment they are emitted, they are absorbed, whereas we see some photons taking up to billions of years to traverse the universe.

"There is nothing in the universe that does not have a reason for it being the way it is. To know a reason for something, we just have to be smart enough to work it out. Of course, there will be reasons that are far beyond the intelligence and knowledge of all of humanity, so we can never expect to know those reasons."

Gregg Thompson

THE STRUCTURE OF MATTER

THE NATURE OF MATTER IN FOUR COMMON STATES

The structure of matter exists in four states – **solids, liquids, gases, and plasmas**. These states are determined largely by temperature, pressure, and the chemical structure of the material. A newly discovered form of matter is a **Bose-Einstein condensate.** It exists near absolute zero (see Volume 1 Chapter 5, page 65). An exotic form of matter known as quark-gluon. Quark-gluon plasma is expected to be the highest level of energy found in Quark Stars. (See Volume 3 Chapter 2, page 65.) Science has discovered other forms of matter that exist under extremely high pressures and densities and also at extreme cold. Under extreme pressure, many gases and ices transform into states with very strange properties.

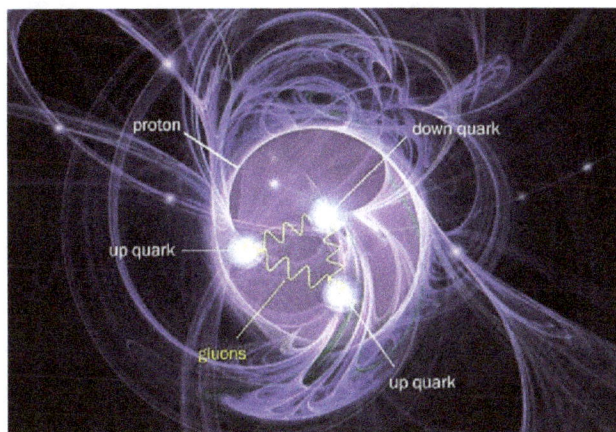

Protons are thought to be made up of a combination of 'up' and 'down' quarks which are held together by a force called gluons.

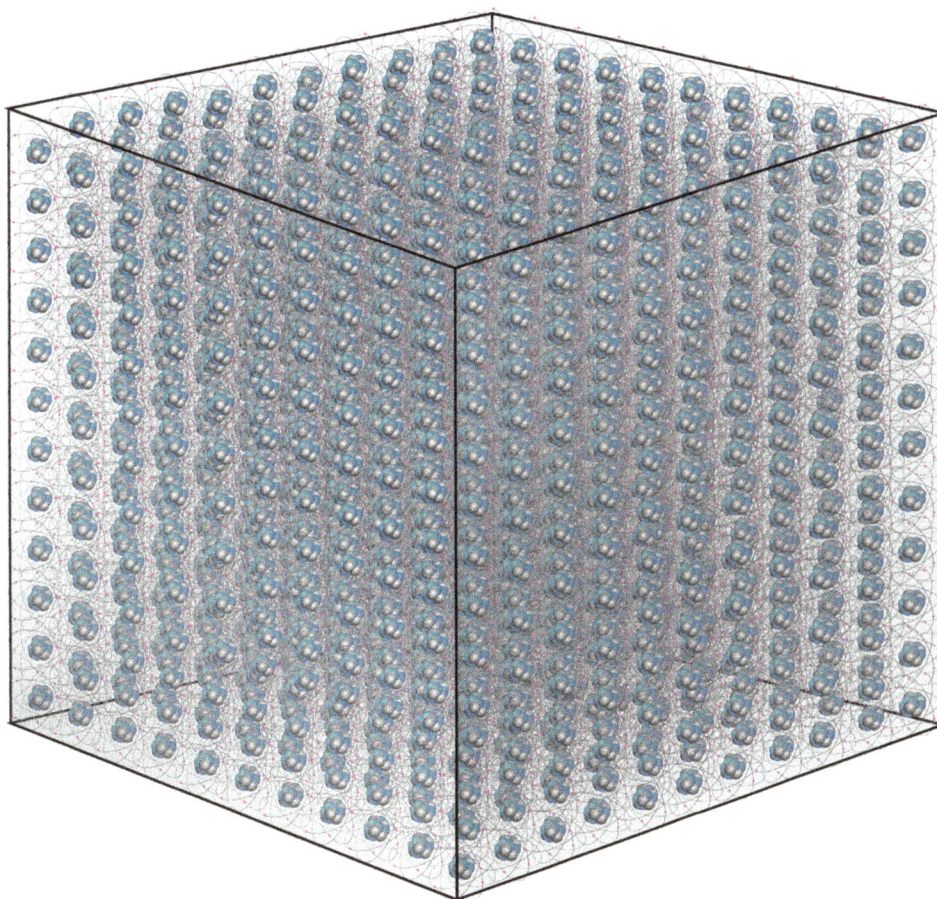

SOLIDS: Atoms in a solid are packed closely together. Their outer electrons orbit adjoining atoms. This locks the atoms together making them typically strong structures. Atoms in solids vibrate but the solid itself does not. When the atoms heat up, their vibrations increase until their electron bonds break, causing the atoms to become a liquid or a gas. Marcus Chown points out in his fascinating book 'Infinity in the Palm of Your Hand' that if all the electrons in a mosquito to explode from their atoms, the nuclei would repel one another,causing and the mosquito to explode with a force equal to a global mass extinction asteroid that would destroy the surface of our planet!

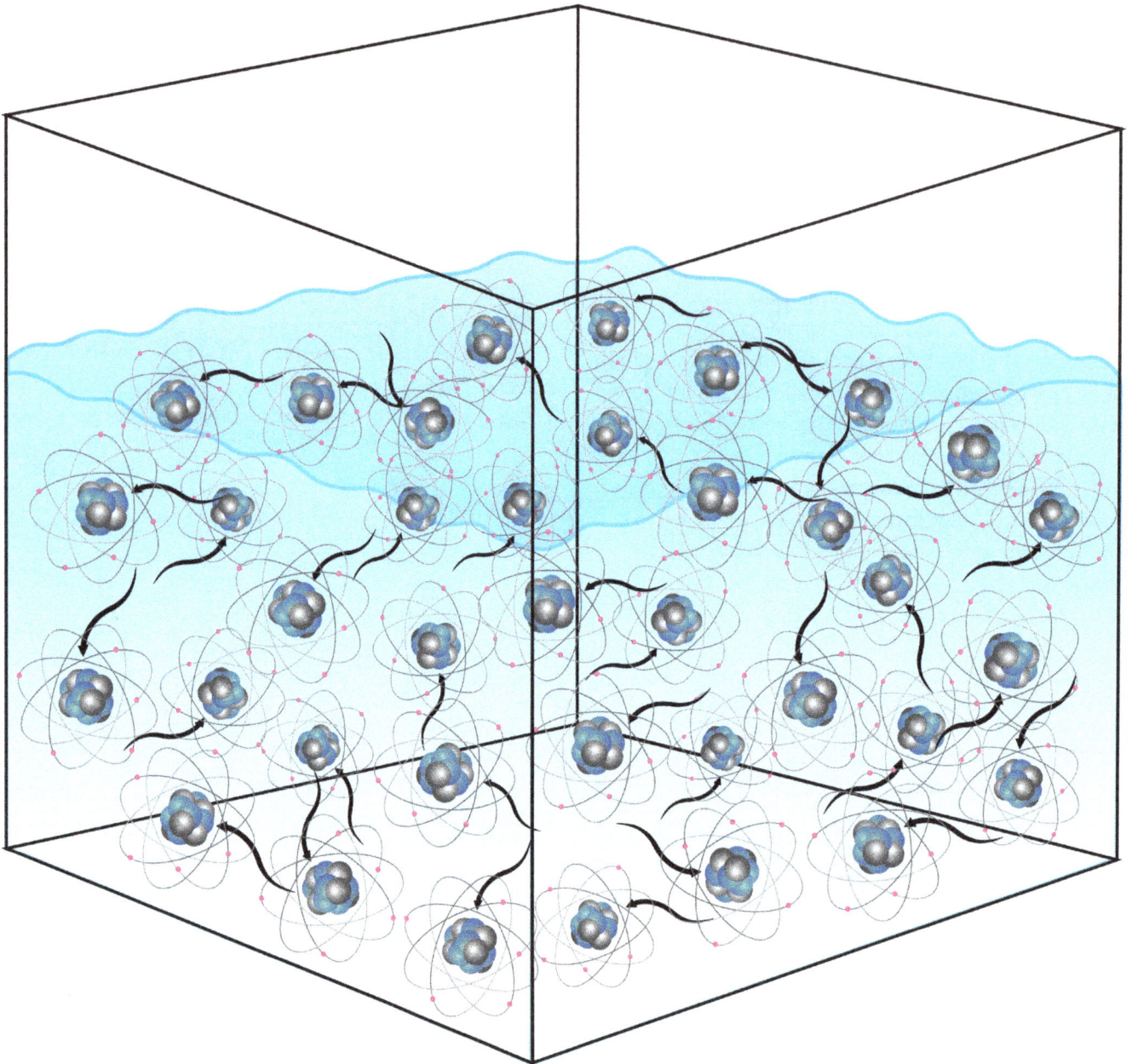

LIQUIDS: Atoms in liquids have enough energy to move around independently. They 'slide' over one another. If they heat up the liquid's boiling point, they have enough energy to fly out of the liquid and evaporate as a gas. Atoms in a liquid are more widely spaced than they are in their solid form.

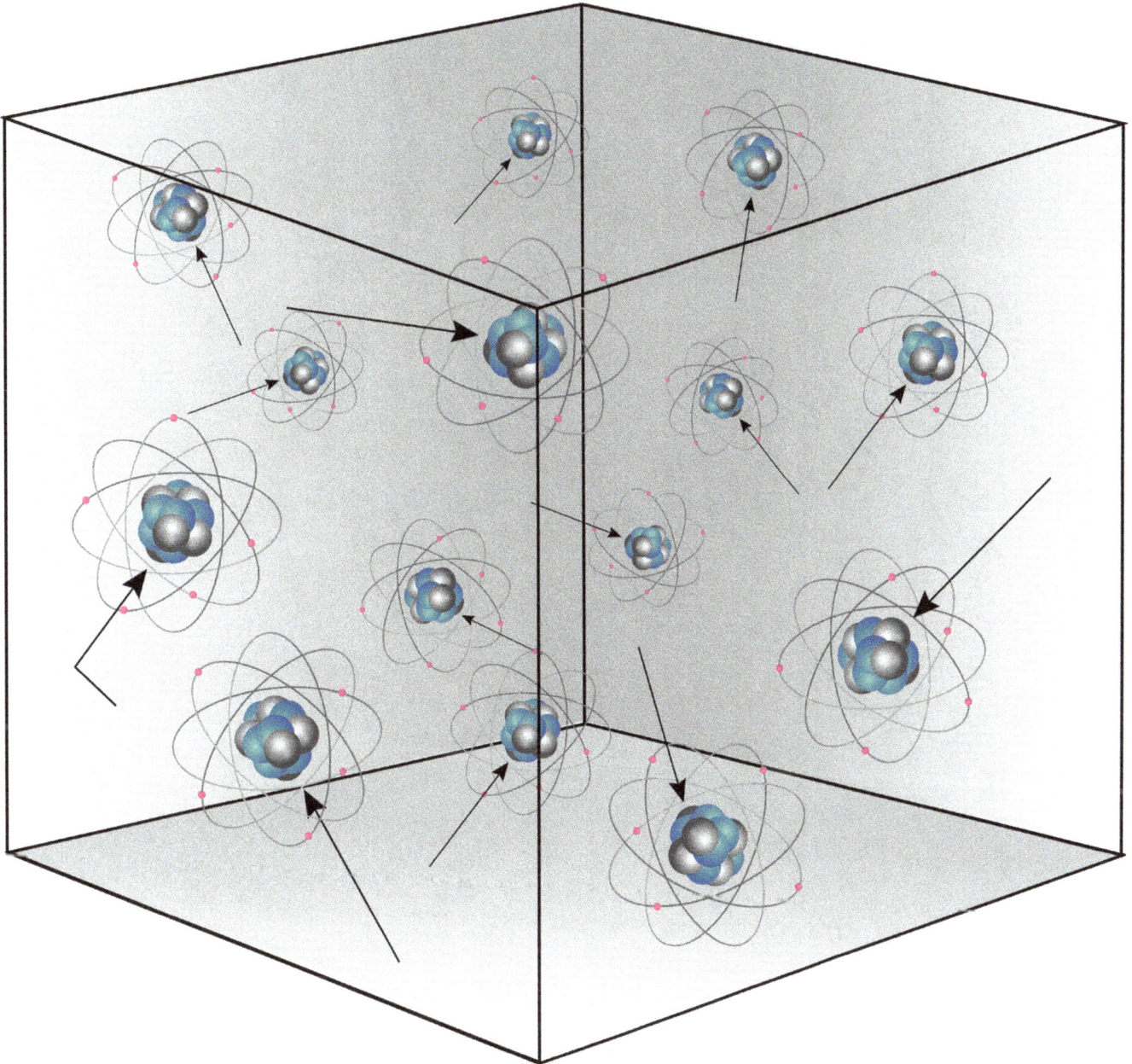

GASES: Atoms in a gas have high energy so they move around at high speeds bouncing off the walls of their container and off one another. If not contained, they dissipate into the atmosphere. Atoms in a gas are far more separated than in a liquid state.

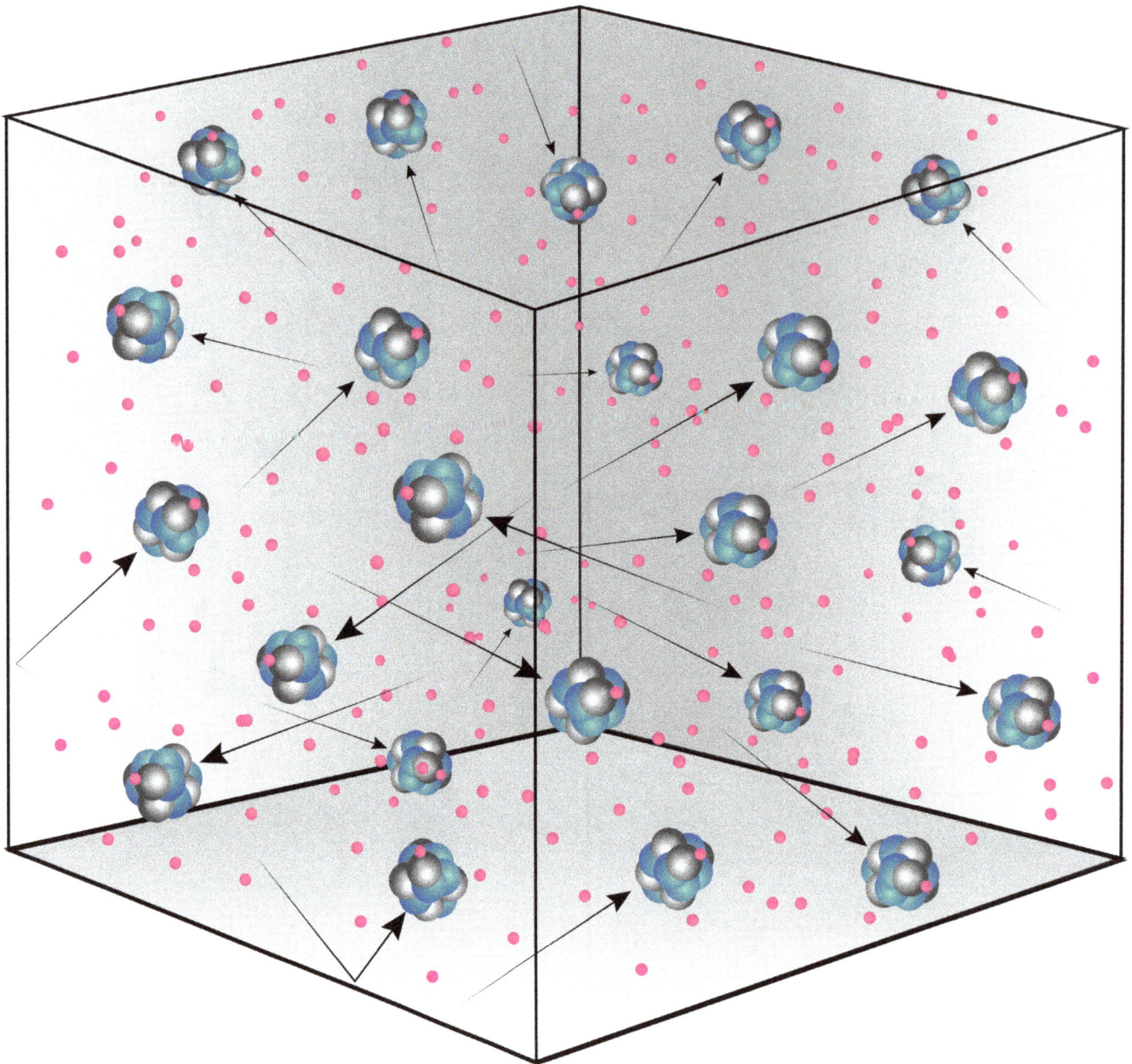

PLASMA: Atoms in a plasma have so much energy that their electrons (pink dots) are stripped away from their atomic nuclei. The nuclei and the electrons move around independently at very high speeds. These sub-atomic particles are widely spaced. Plasmas give off photons of light which are not shown here.

When temperature and pressure continue to rise to extreme levels,, matter changes radically into many other states. See Volume 3 page 42 - Extreme Stars such as neutron stars and supernovas.

AN EXPLANATION OF THE COMMON STATES OF MATTER

We will start with the lowest energy level of matter known as solids and then go to higher energy levels.

In a solid, atoms are rigidly locked together due to their electrons orbiting adjoining atoms. Examples of solids include chalk, metals, ice, and diamonds.

The higher the temperature, the more the atoms in a solid vibrate. When the temperature reaches a solid's melting point, the electron bonds vibrate so much that they break. The atoms are no longer locked together, so they become a liquid. In some cases, they can turn straight into being a gas.

In a liquid, atoms move around, sliding past one another. Examples of liquids are water and oil. When their temperature rises, their atoms move ever faster until they reach their boiling point, at which time some atoms near the surface gain enough energy to fly out of the liquid to become a gas.

In a gas, atoms have enough energy to fly around at fairly high speeds. Their speed is determined by the temperature and the pressure of the gas. Examples of gases are air, water vapor, oxygen, and hydrogen. At high temperatures, electrons separate from the nucleus of gases to form a plasma.

In a plasma, atoms have such a high temperature and therefore high energy that they give off light. The atoms impact each other with such force that their outer electrons, or in some cases all their electrons, are stripped away. This causes the atom to be positively charged. The negatively charged electrons fly around freely. Examples of plasma include a nebula, the Sun's atmosphere, lightning, gas in a neon tube, and a plasma ball.

THE FORCES OF NATURE

Quantum physicists have recently discovered that a vacuum is not nothing: it is filled with the **Higgs field**, which is thought to determine the mass of all particles. The **Large Hadron Collider (LHC)** did not find different kinds of Higgs boson particles as they had expected to: they found just one. It was found to have very simple properties. This suggests that there is extraordinary simplicity in the universe's structure. Science is now discovering that the universe is controlled by simple rules, but when these rules interact, they develop more complex rules, which produce very complex outcomes.

Quantum physicists study the properties of very small particles and the forces that act upon them in the quantum universe. They use the theory of **Quantum Mechanics (QM)** to explain their observations. This theory has been very successful in unifying the three forces of nature, **electromagnetism** (that explains how magnets and electricity work), the **strong nuclear force** (that binds sub-atomic protons and neutrons together inside the nucleus of an atom and which causes nuclear fusion) and the **weak nuclear force** (which is responsible for radioactivity and nuclear fission).

Physicists working in the field of astronomy study very large aspects of the universe, which are controlled by the fourth force of nature - **gravity.** How gravity works is determined by Einstein's **Theory of General Relativity.** Both Quantum Mechanics (QM) and the Theory of Relativity work exceptionally well, but physicists have not been able to combine gravity with the other three forces so that there is one unifying theory that explains everything. Einstein spent the latter half of his life trying to do this without success. There is sure to be something that has not yet been identified that will solve this problem.

A BRIEF SUMMARY OF STRING THEORY

String Theory will only be able to unify the four forces of nature as long as there are *nine* physical dimensions: not just the three dimensions that we perceive. This theory offers some explanations for some aspects of QM, but physicists have not found a way to prove that more than three physical dimensions exist. Maybe in the future, highly advanced computers will discover a way to prove that other dimensions do exist. If other dimensions do not exist, then String Theory is simply a wonderful, elegant theory that is entirely wrong.

String Theory suggests that the smallest elements that make up the universe are incredibly tiny 'strings' of energy that vibrate in all sorts of patterns to create the huge diversity of particles and forces found in the universe.

Cosmologist **Leonard Susskind** cleverly conceived String Theory to explain the workings of QM and to hopefully unify quantum gravity with the other three forces. He imagined that the most basic elements in nature consist of infinitesimally small energy packages (a million, billion times smaller than the smallest atom, hydrogen) in the form of 'strings' that are either open lengths or closed loops i.e. strings; hence the name of his theory. These strings vibrate in numerous ways to give each type of quark different properties. Susskind cannot explain what makes his strings vibrate. (To me it sounds like a really elegant concept, but one which is not real. It reminds me of the elegant crystal spheres concept that explained the sizes of the orbits of the planets. But String Theory was very popular amongst quantum physicists to the point that if you don't work in this field, you are banished.) But this is now changing.

String Theory suggests that the smallest elements that make up the universe are incredibly tiny 'strings' of energy that vibrate in all sorts of patterns to create the huge diversity of particles and forces that are found in the universe.

Susskind envisaged strings as being one-dimensional lengths of *energy*, whereas QM sees particles of matter as zero-dimensional *points* where each particle vibrates in its own unique way, similar to how strings in musical instruments vibrate. Various combinations of these strings would create all the particles of matter. Low energy vibrations produce photons and high energy vibrations to create massive particles. The number of varieties of strings (if they exist) have not been calculated. In QM there are over a thousand forms of matter made up of quarks. Susskind uses an analogy from biology to explain how all the different types of particles may be controlled by a basic form of energy. He recounts how biologists could not work out how there could be such an incredible diversity of life forms, ranging from bacteria all the way up to humans - until they probed deeper and deeper into the smallest parts of organic matter and eventually found the answer - DNA! They discovered that **DNA's double helix** had four different molecular bars, which are commonly referred to as 'letters'. These chemical 'letters' that are arranged in all sorts of combinations to construct a near-infinite diversity of genes. These became the design codes for innumerable life forms. Susskind theorizes that, like the letters of DNA, his stretchy vibrating strings form the basic codes of matter that determine the properties of each particle.

Strings are inconceivably small. If we blew up a tiny atom to the size of the observable universe, then a string would only be as large as an apple tree! If Susskind is on the right track, then strings might be the most basic element from which everything is constructed. String theorist **Lee Smolin** in his book '*The Trouble with Physics*', provides numerous reasons why String Theory is now unlikely to be able to describe reality.

Some big problems with String Theory are that it has almost no mathematical restraints so virtually anything is possible! That being the case, one possibility must be that strings do not exist. It also leads to an infinite diversity of universes, but when infinites are involved, this too is typically indicative of errors. A theory having unrestrained infinite possibilities makes it useless for making predictions because it allows anything to happen so there is no way to prove it to be either true or false. For a theory to be of any value, it must be able to make predictions that can be tested. If String Theory really is a good approximation of reality, then we must be overlooking some fundamental aspect of it that would get rid of the infinities involved and its lack of restraints. It is so complex that very few physicists understand it.

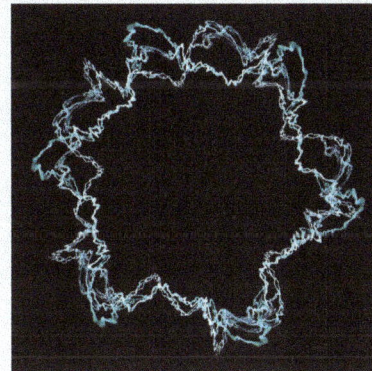

Theorists visualize strings as being vibrating energy fields something like this.

CHAPTER 3

SUPPORT FOR THE BIG BANG THEORY

At the instant of the Big Bang, it is thought that the universe had near infinite temperature and near infinite density. For an instant, it was all energy (colored wisps) but as it expanded and cooled, particles of matter started to form (white spots). At a billionth of a billionth of a second, every part of space experienced an extraordinarily rapid inflation causing it to expand much faster than the speed of light for a mere trillionth of a second. After that, it expanded at the rate we observe today, which is much slower but still faster than light. The Big Bang did not expand into a pre-existing space: it caused space itself to expand. As it did, its temperature was dispersed, so it cooled. Around half a million years later, the first stars started to form. Over a billion years or so later, increasingly complex chemistry started to evolve on some planets with the right conditions to produce the first forms of life.

When air cools, mist globules in clouds combine to become ever larger and heavier until they cannot float in the air any longer so they fall as raindrops, hail or snow. Similarly, as the near infinite temperature of the Big Bang started to cool due to the universe expanding and spreading its temperature out, the most basic particles that we know of, quarks, clung together to form heavier particles such as protons and neutrons.

If air cools rapidly to below freezing point, rain droplets experience a phase change and become ice crystals. When they combine and grow large enough, they fall as snow or hailstones. Similarly, as the Big Bang continued to cool, its protons, neutrons, and electrons came together to form atoms. The rain drop-hail analogy suggests how the early universe appears to have through phase changes in its states of matter.

It's important to understand that the Big Bang was not like a bomb on Earth exploding into some pre-existing volume of space. Rather, space itself expanded in volume *at every point across the universe carrying matter with it -* and space is still expanding but its rate of expansion has been increasing for the blast few billion years. (The expansion of space concept goes against all that humans are aware of in their everyday experiences, so if you find it difficult to visualize this, then you are not alone.) Matter does not expand, only space does.

The Big Bang theory offers no explanation as to where its energy came from. The Big Bang theory is about the evolution of the universe: it makes no attempt to answer what produced the energy for the creation of the universe. Nor does it say anything about there being an 'outside' to the universe. Asking what is outside the universe is like asking what is north of the North Pole. It is meaningless. An outside to the universe is nothing more than a philosophical concept that has no way of being proven true or false. Such speculation is outside the realm of physics, so it is not science.

The Big Bang theory is strongly supported by observational evidence, as shown below. **The First Law Thermodynamics** states that matter and energy cannot be created, nor can it be destroyed. It can, however, change from one form to another. There are some who say that because the Big Bang appears to have created matter and energy, it violates the first law. But the Big Bang theory does not describe where the energy came from because the laws of physics break down close to the birth of the universe due to the temperature and pressure being virtually infinite. Under these conditions, there is surely a state of physics but one which we are not yet aware of.

For any theory to be valid, it must be able to make predictions and there must be strong evidence. to support it. This is the case with the Big Bang theory. Before the Big Bang theory was developed, it was universally believed that the universe was eternal and static, but all the evidence that supported the Big Bang proved that the universe was not static and eternal; It was very dynamic.

The Big Bang theory predicted that:

1. **The universe would have a birth, and it would age.** Einstein believed in a static universe until 1917 when he realized that his General Theory of Relativity predicted an ever-evolving universe. In the late 1920s and early 30s, astronomer **Edwin Hubble** made red shift observations of galaxies that proved that the universe was not static because galaxies were moving away from one another. The only explanation for this was that the universe was expanding. If a video of this was run in reverse, then it would show that all the matter in the universe must have come from a single point! This was a truly astonishing, mind-bending concept, especially for those times. Hubble's observations proved that the **Steady State Theory** of a static universe, which was promoted by **Fred Hoyle, Thomas Gold** and **Hermann Bondi**, could not be true.

Ironically, it was physicists Fred Hoyle who brilliantly worked out how stars produced energy through nuclear fusion. This meant that stars would have to be born and that they would eventually burn out and die, so the universe could not be static as Hoyle believed. The largest stars would burn the hottest and fastest, so they would be short-lived, and the smaller a star was, the slower it would burn, so the longer it would live. By proving how stars produced their high energy output, Hoyle proved his own Steady State theory had to be wrong! But he doggedly refused to accept that.

Because Edwin Hubble had initially miscalculated the age of the universe to be less than that the oldest rocks on Earth, Fred Hoyle, in an attempt to discredit the Big Bang theory, used Hubble's mistake to try to disprove the Big Bang. But Hubble soon realized that he had made a mistake about the brightness of the types of stars he used for distance measurements to other galaxies. When he corrected for this, he saw that the universe was indeed older than the oldest rocks on Earth. (See Volume 3 Chapter 12, page 285 for details.)

We now know that everything in the universe is evolving. We know that stars and planets are constantly being born and others dying and that galaxies continually evolve and merge to form giant galaxies. They then group together to form superclusters of galaxies. Observations of the most distant galaxies at the edge of the observable universe are seen as being very young as they were at the birth of the universe. They have young, newly-born stars. This proves that the universe had a birth. In closer galaxies, we see many stars that are very old as well as newly-born stars in nebulas, so the prediction that the universe had a birth, evolved and now has old stars, is true.

2. **The Big Bang predicted that matter in the universe would be uniformly distributed on large scales.** This was proven to be the case because Edwin Hubble was able to show that the universe looked the same in every direction on a large scale. Hubble Space Telescope images have demonstrated this beautifully.

3. **The Big Bang predicted that the universe would be expanding.** Einstein's calculations predicted that the universe should expand but this sounded so preposterous to him that he rejected it. To overcome this, he introduced a 'cosmological constant' into his equations to cancel out the expansion. When he later learned that Hubble's redshift observations of galaxies

showed that the universe was indeed expanding, he said this was his greatest mistake.

Georges Lemaitre

In 1923 as a graduate student, **Georges Lemaitre** was introduced to cosmology by the great astronomer **Arthur Eddington**, and a year later while at Harvard College Observatory in Cambridge, Lemaitre spent a year working with another great astronomer, **Harlow Shapley**. In 1925, Lemaitre had become a Belgian physicist and a priest when he started to develop his theory that the universe was expanding from a "primeval atom". This was prior to Edwin Hubble's observations of the redshifts of galaxies that showed they were receding from us. Two years later, he was sure his theory was right after he combined Einstein's theory of General Relativity with Hubble's newly made observations. Lemaitre tied to explain this to Einstein, but he would not listen. Einstein refused to abandon the static universe! But in 1931, Einstein finally adopted the expanding universe model because the evidence for it had become widely accepted. He abandoned his cosmological constant that canceled out his math that predicted an expanding universe. The belief that the universe was unchanging came from the genius **Sir Isaac Newton** in the 1700s. He assumed that God made the Earth and the universe unchanging and everlasting, and that stars were motionless and ever-lasting. At that time, this seemed realistic and obvious because the universe looked that way. This concept was unquestionably accepted as fact for 200 years, because in those days there was a total lack of any opposing evidence. Had there been evidence to the contrary, Newton would probably have changed his mind.

The Redshift of Galaxies: When light from galaxies was passed through a prism to split it up into its constituent colors, known as its spectrum, it was discovered that the colors were shifted towards the red end of the spectrum. The further away a galaxy was, the more its spectrum was shifted to the red end. This is due to the Doppler Effect which causes light waves to be stretched if an object is moving away from the observer. This makes them appear redder. If an object is moving towards the observer the light waves are compressed, making them appear bluer and they move towards the blue end. The degree to which a galaxy's light is red-shifted determines how fast it is receding from us, and similarly, the degree to which it is blue shifted is a measure of how fast it is approaching us. It was seen that the more distant a galaxy was, the more it was red-shifted. The only explanation for this was that the universe must be expanding even faster as time goes on. Only a few nearby dwarf galaxies and the Andromeda galaxy have blue shifts. All other galaxies have redshifts, so they are moving away from us due to the expansion of the universe.

THE DOPPLER EFFECT ON LIGHT

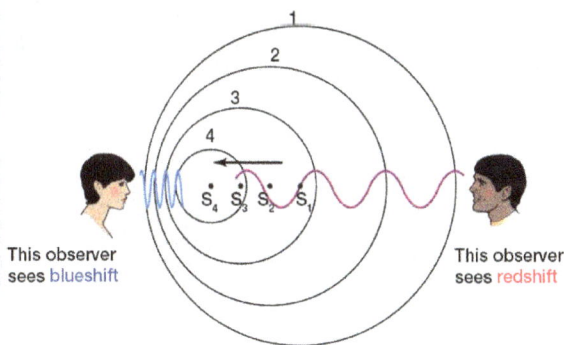

The dots moving to the left representing light waves moving towards the woman observer are compressed making the light waves appear bluer, while those on the right are moving away from the male observer lengthening thereby making them appear redder. The faster an object is moving, the greater the effect is. The same happens with sound: the pitch of the sound of an ambulance coming towards us rises and when it goes away from us, the pitch becomes lower.

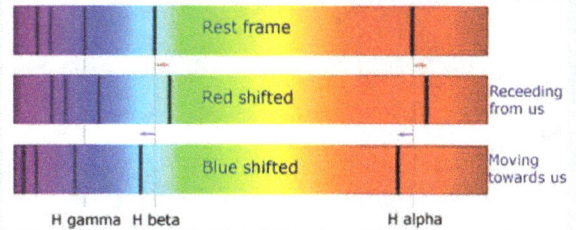

When an object like a galaxy is moving away from us, its spectrum becomes increasingly redder the faster it is receding from us. This is known as its redshift. An object moving towards us appears bluer.

4. **The Big Bang predicted that the expansion rate would increase over time due to an anti-gravity force.** This seemed impossible until it was recently discovered through observing distant supernovas that the rate of the expansion of the universe appeared to be increasing. It was also discovered that there appeared to be an anti-gravity force in the center of the large voids between the **filamentary structure** throughout the universe. This 'force' became known as **dark energy**. It appears to repel matter, and therefore galaxies. (For details on dark energy, see page 37, and for the Filamentary Structure see Chapter 13.) It is thought that dark energy pushes galaxies out of the voids to their outer edges forcing galaxies to congregate along the filaments.

5. **The Big Bang predicted that because the expansion of space throughout the greater universe occurred everywhere at the same time, radiation from it should be distributed evenly throughout the universe. It should be observed as a microwave background with a temperature only 3 degrees above absolute zero.** Both the background radiation and its exact temperature were first confirmed by satellite observations in 1989, and again with another more advanced satellite in 2001. To have such a perfect match between theory and observation was very well received by cosmologists. This was yet another important proof for the Big Bang Theory.

HOW DID OUR UNIVERSE BEGIN?

Some 13.8 billion years ago our entire visible universe was contained in an unimaginably hot, dense point, a billionth the size of a nuclear particle. Since then it has expanded—a lot—fighting gravity all the way.

Inflation
In far less than a nanosecond a repulsive energy field inflates space to visible size and fills it with a soup of subatomic particles called quarks.

Age: 10^{32} milliseconds
Size: Infinitesimal to golf ball

Early building blocks
The universe expands, cools. Quarks clump into protons and neutrons, the building blocks of atomic nuclei. Perhaps dark matter forms.

.01 milliseconds
0.1-trillionth present size

First nuclei
As the universe continues to cool, the lightest nuclei, of hydrogen and helium, arise. A thick fog of particles blocks all light.

.01 to 200 seconds
1-billionth present size

First
As ele
nuclei
from
unveil
as ou

380,
.000

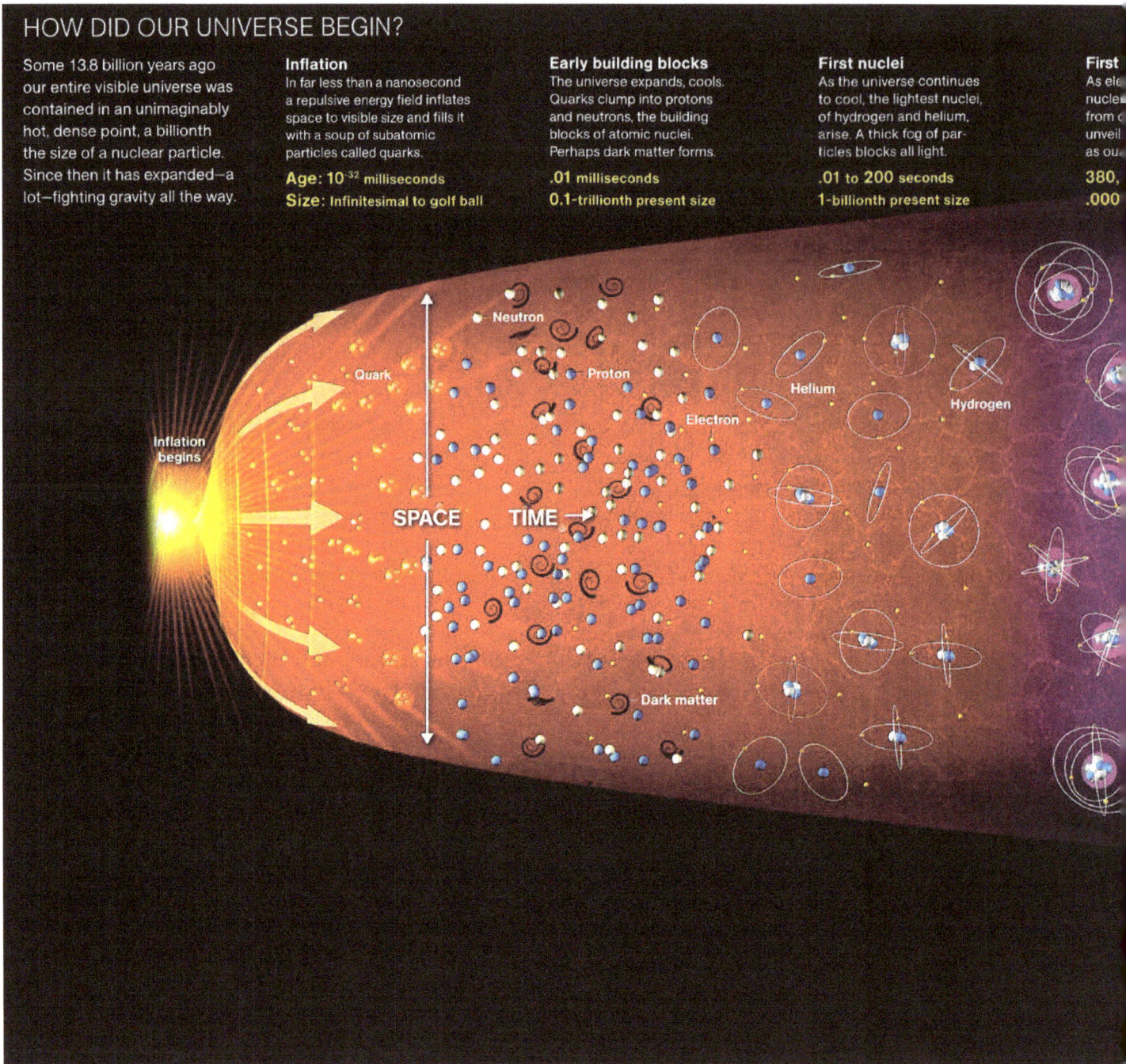

Labels: Neutron, Quark, Proton, Helium, Hydrogen, Electron, Inflation begins, SPACE, TIME, Dark matter

The Big Bang was the beginning of the evolution of ever-increasing complexity in the universe.

From left to right: It is thought that the four basic subatomic particles to form were photons, negatively charged electrons, positively charged protons, and neutrally charged neutrons. (It is thought that there were also equal numbers of anti particles but for an unknown reason, only normal matter survived.) As the universe cooled, protons and neutrons combined to form the nuclei of atoms. When the universe cooled sufficiently, negatively charged electrons orbited positively- charged atomic nuclei to form neutrally charged atoms. Atoms condensed into gas and dust to eventually form stars and their planets. Gravity drew stars together to form galaxies. Large stars exploded to form all the heavy elements, which enabled planets to form and life to evolve. Credit: National Geographic

ht
ting
the glow
e is
far back
see.

The "dark ages"
For 300 million years this
cosmic background radiation
is the only light. Clumps of
matter that will become
galaxies glow brightest.

380,000 to 300 million years
.0009 to 0.1 present size

Gravity wins: first stars
Dense gas clouds collapse
under their own gravity—and that
of dark matter—to eventually
form galaxies and stars. Nuclear
fusion lights up the stars.

300 million years
0.1 present size

Antigravity wins
After being slowed for
billions of years by gravity,
cosmic expansion acceler-
ates again. The culprit: dark
energy. Its nature: unclear.

10 billion years
.77 present size

Today
The universe continues
to expand, becoming
ever less dense. As a
result, fewer new stars
and galaxies are forming.

13.8 billion years
Present size

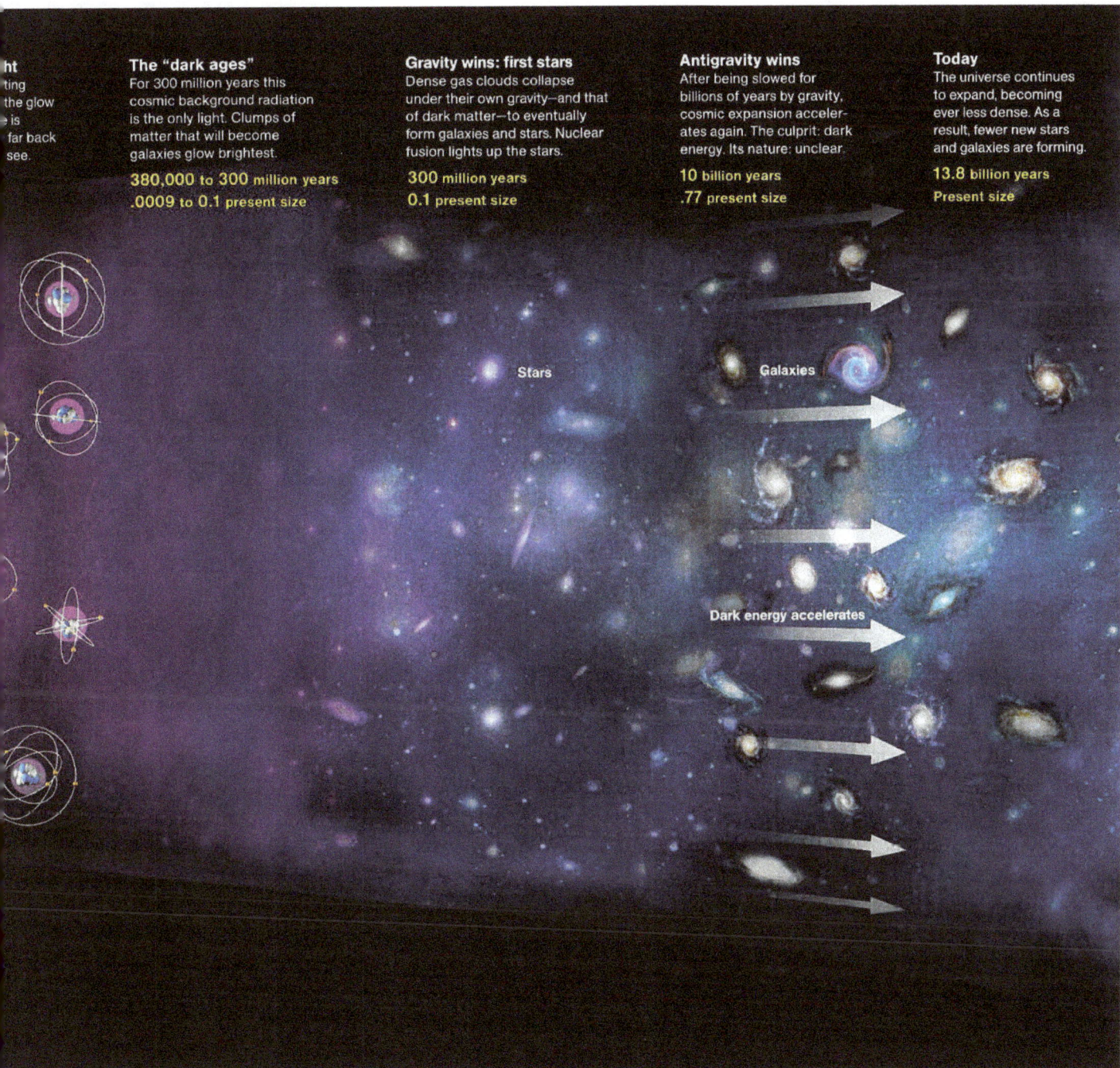

Stars

Galaxies

Dark energy accelerates

6. **The Big Bang predicted that the background of the universe would be black.** If the universe was transparent, infinite and unchanging, then the night sky would be white hot, like the surface of a star. (See **Obler's Paradox** page 31.) There would therefore be no planets and no life. The light and heat of the innumerable distant galaxies in the Greater Universe would fill every gap between the galaxies that we can see making the sky intensely bright and hot everywhere. But the Big Bang theory is underpinned by the **Theory of Relativity**, which places a limit on the speed of light. This causes the part of the universe that we can observe to have a boundary. Due to this, the further we look, the further we look back in time until we eventually see back to the beginning of the universe – the beginning of time itself! This means that **we can only see a finite volume of space.** This volume is known as the '**observable universe**'. (See Part 4.) As the universe we observe is not infinite, this means that the background sky must be black because there is not an infinite number of galaxies visible to fill every part of the sky.

After almost a century of astronomers and physicists trying to find holes in the Big Bang Theory, nearly all cosmologists, astronomers, and physicists now believe the evidence for the Big Bang occurring is so overwhelming that it is well accepted. There is no other theory that can explain the evidence that supports the Big Bang theory. However, as we learn more, we are likely to find that certain aspects of this theory may not be as we assume them to be today. This occurs with most theories.

HOW THE UNIVERSE EVOLVED

To easily understand the early stages of the Big Bang requires some knowledge of the basics of physics and chemistry. For readers who are unfamiliar with these sciences, this section provides an overview using everyday language to explain the main stages of the universe's evolution. While most aspects in the early phases of the Big Bang are considered fairly solid, the detailed physics underpinning the first quadrillionths of a millisecond is somewhat speculative. Exactly what physicists think occurred then will be far better understood in the coming couple of decades when mega-intelligent computers and enormous telescopes can give us great insight into things we know nothing about today. Of critical importance is the fact that it is not yet known what caused the Big Bang to occur, or where its infinite energy came from.

To develop the following history of the universe required a great deal of work over the last century by cosmologists, astronomers, physicists, and mathematicians. With a lack of data in some areas, they have been forced to add a dash of speculation here and there. This is evident by the variation in figures provided from one source to another, especially in the very early stages of the Big Bang.

Based on what we know today about the physics of the Big Bang and how life evolved on Earth, the following table is a simplified chronology of the universe's evolution.

CHRONOLOGY OF THE UNIVERSE IN BILLIONS OF YEARS

ELAPSED BILLIONS OF YEARS SINCE BIG BANG	BILLIONS OF YEARS AGO	COMMENTS
0	13.8	Big Bang
0.4	13.3	Matter & light decouple
0.6	13.1	First stars formed
0.8	12.9	First supernovas occurred
1.2	12.7	First galaxies formed
1.5	12.3	First planets formed
1.8	12.0	Simple molecules form then complex organic chemistry commences
2.1	11.7	Building blocks of life evolved
3.6	10.2	First cells evolve on suitable planets
6.1	7.7	First advanced multi-cellular life equivalent to dinosaurs evolved
6.3	7.5	First intelligent biological life forms create non-biological computer-based super minds
9.1	4.7	Our solar system formed
13.8	Now	Humans on Earth create first non-biological, computer-based mega-intelligence. (This most likely occurred numerous times across the Greater Universe up to 9 billion years before it occurred on Earth.) Humanity starts to become aware of the role that intelligence and consciousness plays in the evolution of the universe.

EPOCHS OF THE BIG BANG

The following is a more detailed account of how the Big Bang unfolded for those who like to understand the basic physics and the time frames involved.

Soon after the Big Bang occurred, it is thought that the temperature was almost infinitely hot. All the energy was compressed in four force fields– gravity, electromagnetism, and the weak and strong nuclear forces. **At 10^{-43} sec** (that's a decimal point with 43 zeros after it and 1 at the end!) it is thought that a phase change occurred, causing gravity to separate from the other forces which remained unified. At this time, the universe was only 10^{-30} times the size of a proton! Quarks and their antiparticles are thought to have been created then.

At **10^{-35} of a second** after the Big Bang occurred, the temperature had dropped to 10^{27} °C - that's 10 billion, trillion, trillion degrees. At 10^{-32} sec, the strong nuclear force separated from the weak nuclear and the electromagnetic force which remained unified. The separating of the Strong Nuclear force is thought to have triggered an incredibly rapid exponential expansion of the universe known as the **cosmic inflation**. The universe expanded almost instantly by 10^{26} times to around the size of a grain of sand or perhaps a grapefruit. This distributed the elementary particles evenly and thinly across the tiny universe.

In the 1980s, cosmologists **Alan Guth** and **Andrei Linde** proposed the **Theory of Inflation** which made our infinitesimally small universe expand extremely rapidly. This might not sound like much of an increase if you don't realize how small a proton is, but it was an astonishing expansion in such a small timeframe. Space doubled in size every trillionth, trillionth, trillionth of a second! We do not know for sure that this occurred, but it explains many aspects of the universe that cannot be explained any other way.

It is not yet known why the universe must have expanded by an inconceivable 10^{50} times in such a short period, and then suddenly, for no apparent reason, slowed to the more moderate expansion rate we see today. Inflation has its critics, but it has been widely accepted because, without it, cosmologists cannot explain the uniform distribution of matter across the universe, which only

varies by a mere 1 part in 100,000. This was discovered by the **Cosmic Microwave Background (CMB)** radiation map of the universe taken by the **COBE satellite** in 1989. In 2001, the **Wilkinson Microwave Anisotropy Probe (WMAP)** measured this in much finer detail. The small variations in the background temperature are thought to have occurred from low amplitude quantum fluctuations during the outset of the Big Bang. What caused these fluctuations is unknown. They could only have occurred if there was more matter in the universe than we see today. Invisible dark matter and dark energy could explain these slight variations. If the distribution of the Big Bang's energy had been perfectly even, then all galaxies would be equidistant from one another so there would be no galaxy clusters and no filamentary structure to the universe. (See Volume 1 Chapter 2, page 17). Between 2009 and 2013, the European Planck cosmology probe made an all-sky map of the cosmic microwave background in higher resolution than WMAP. According to the Plank satellite's map, the subtle fluctuations in temperature were imprinted on the deep sky when the universe was only about 370,000 years old. It arose as early as the first nonillionth (1 followed by 30 zeros) after the Big Bang occurred. It is currently th eorized that these fluctuations gave rise to the present vast cosmic web of galaxies and the voids between them that maybe filled with dark matter.

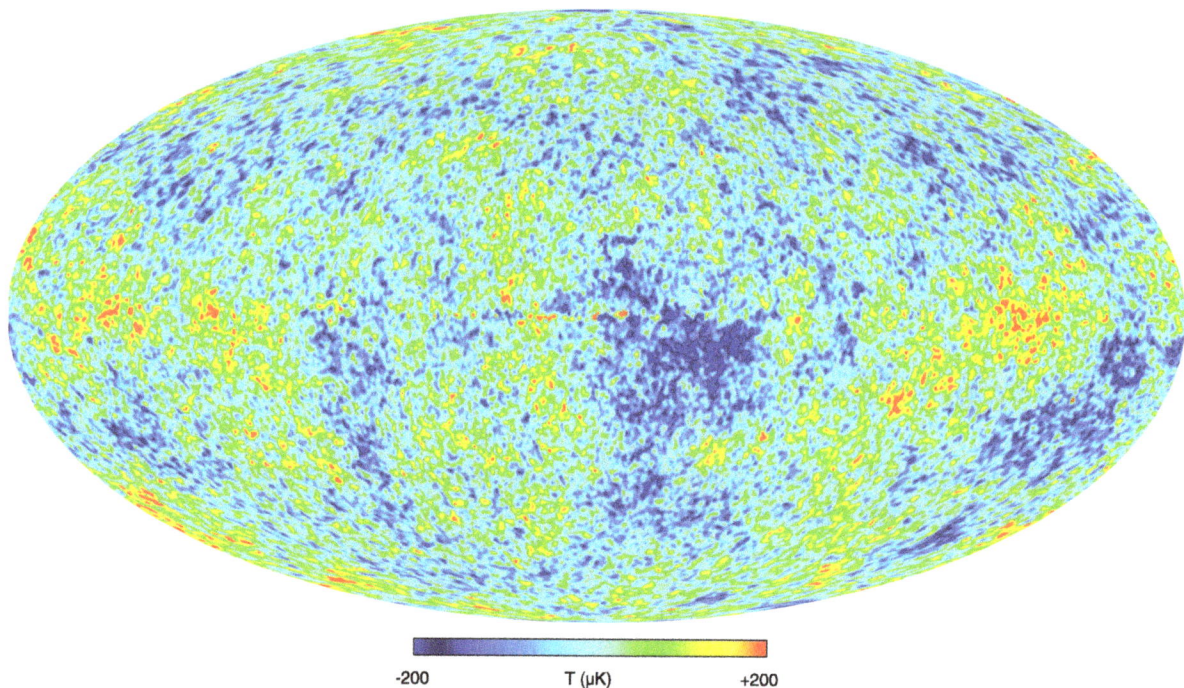

-200   T (μK)   +200

The Cosmic Microwave Radiation Background map shows extremely small variations in temperature in the universe shortly after the Big Bang occurred. Bright red regions are ever so slightly warmer than the colder dark blue regions.

Science is realizing that imperfections are necessary at every level of the universe to make it evolve. Without these imperfections, the universe would be sterile. (A detailed explanation of this is given on page 71.)

At this time, there was a soup of quarks - some of which *may* have given rise to dark matter.

At 10^{-5} **of a second**, the temperature dropped to a point where quarks could bind together to form particles such as protons and neutrons.

For another unknown reason, at 10^{-4} **of a second** ($\frac{1}{10,000}$th of a second) after the Big Bang, when the temperature had dropped to 10^{13} °C (1 hundred, thousand, billion degrees), there appeared to have been a slightly smaller production rate for **anti-matter** than there was for ordinary matter. We do not yet know what could have caused this to occur, but if **there was <u>not</u> an imbalance, the universe would have been devoid of matter** because antimatter and matter annihilate one another upon contact. The **imbalance between anti-matter and normal-matter** resulted in normal matter dominating the universe. For every 100 million anti-protons formed, there was just one extra proton of normal matter created and that was enough to form all that we see in our observable universe and the Greater Universe beyond!

When electrons collide with anti-matter electrons (called positrons), they annihilate each other producing an abundance of massless photons (light). **For every proton or neutron in the universe, there were about 2 billion times as many photons produced. That produced a lot of light!** However, photons could not travel very far without hitting another particle that had mass so they were therefore absorbed. This resulted in the universe being dark and opaque for 380,000 years after the Big Bang.

Physicists calculate that at **1 second** after the Big Bang, the density of matter was so high that positively charged protons and negatively charged electrons were crushed together to form neutrons. At **3 seconds**, there were free protons that became hydrogen nuclei. The pressure was high enough then for some neutrons to combine with protons to convert 3% of the hydrogen into **helium** as well as a very small amount of **lithium**.

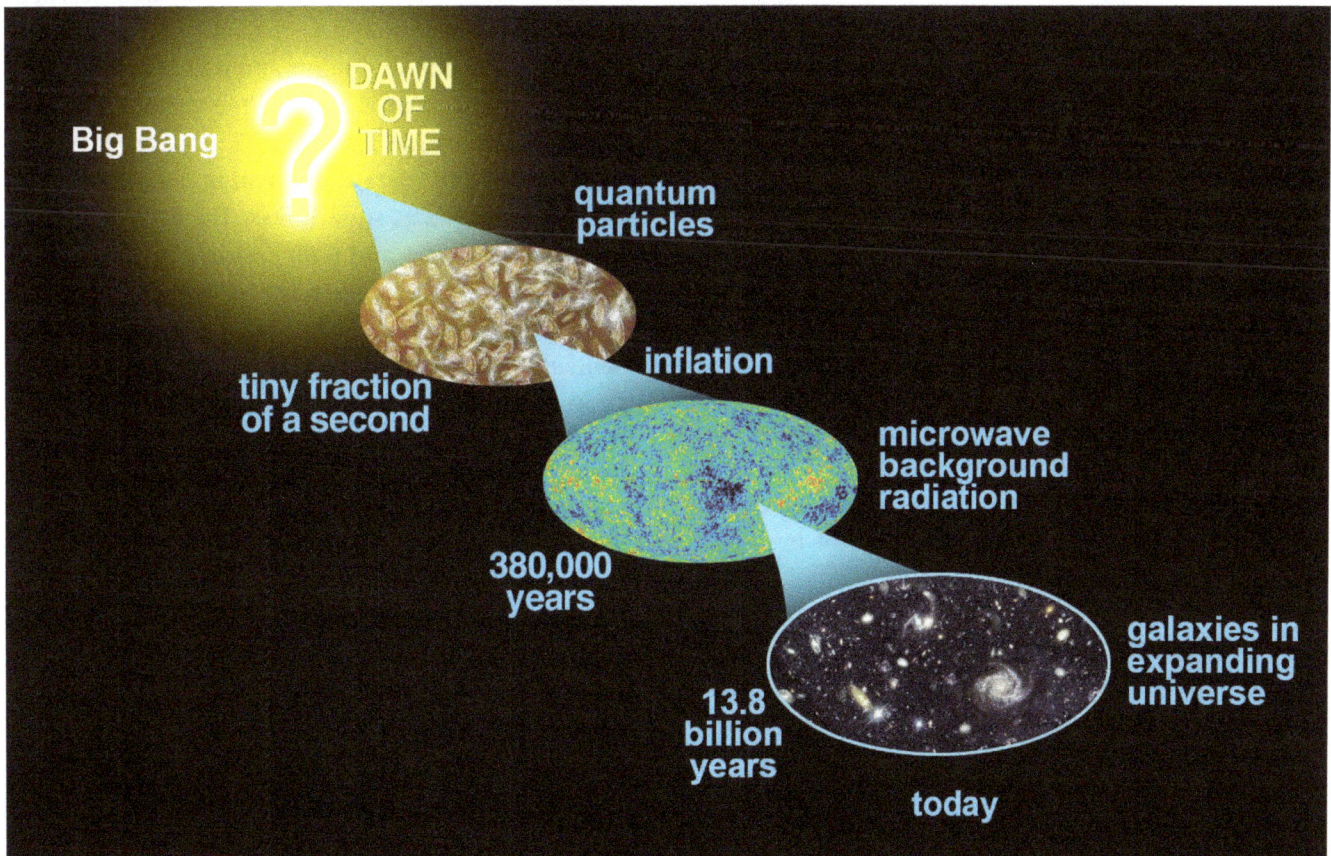

The three main stages in the expansion of the universe Credit: NASA

Around **100 seconds** after the Big Bang commenced, expansion dropped the temperature to **109 °C** (1 billion degrees). At that point, the universe was a dense soup of particles entangled with photons. The universe was far too dense for light to be able to travel far before impacting nearby particles. The cosmic background radiation map (see page 24) gives us a temperature picture of what the universe looked like at this very early time.

If we jump ahead to around **56,000 years**, the density of matter equaled the density of all forms of radiation energy. The temperature had now dropped to **9,000°C**, but it was still too hot to form stable, neutral atoms because there were too many collisions between photons, atomic nuclei of protons, and neutrons, for electrons to be able to maintain an orbit around atomic nuclei.

The universe was a very compact, opaque plasma of particles.

At around **380,000 years,** the universe had expanded enough to have cooled to 3,000°C. This marked the beginning of the **end of the dark era.** With the continuing expansion of the universe, the temperature dropped enough to allow positively charge nuclei of hydrogen and helium to capture negatively charged electrons to form the first **neutral atoms.** At this time, the mass of the universe was approximately 75% hydrogen, 24% helium with less than 1% lithium. These were the only elements in the universe that formed normal matter. (This ignores dark matter, if it exists.)

Around 500,000 years, the universe's opaque plasma had expanded and cooled enough to allow particles and photons to move far enough apart so that collisions were now becoming infrequent. Much of the energy at this time was in the form of invisible microwave radiation.

An afterglow of the Big Bang still remains and satellites have recorded this as the microwave background radiation.

With the universe's continuing expansion, light was now able to travel increasingly further, thereby making the universe become increasingly more transparent. Light produced during the Big Bang was a dull background glow that could now travel further across space. But **the universe was still mostly dark because there were no stars at this time.** It is estimated that it probably took around a billion years for the universe to expand enough for most of the dark fog from the creation of the universe to fully dissipate to make the universe clear like it is today.

The universe remained relatively dark until around **200 to 400 million years** after the Big Bang at which time gravity began to cause matter and dark matter to clump together. Where there were slightly denser regions of it, the first stars and planets formed. **Around 500 million years, the universe was now lit by trillions and trillions of newly formed stars**. By **a billion years,** gravity had drawn these stars together to form galaxies.

High temperatures and pressures in the cores of the stars 'cooked up' the light elements of hydrogen and helium to form heavier elements up to iron. These included the most important ones such as carbon, nitrogen, oxygen, sodium, magnesium, aluminum, silicon phosphorus, and sulfur and well as several other less common elements. Refer to the Periodic Table of Elements in Volume 3 Chapter 2, page 72.) **Within just a few million years after forming, the largest stars started exploding as supernovas. The extremely high gravitational forces and temperatures in the cores of supernovas fused some the lighter elements together to create traces of all the other 70 odd heavier elements up to uranium.** These heavier elements became dispersed throughout galaxies and then later incorporated in future generations of stars and their planets. **Without these heavier elements, the most complex chemistry of life could not have evolved.**

At **1.2 billion years,** irregular galaxies started to condense enough to make them rotate, thereby forming spiral galaxies. Galaxies that were relatively close together were attracted to one another by their mutual gravity. They formed **the first groups of galaxies**, which then combined with others to form **clusters of galaxies.**

At **1.8 billion years, simple molecules formed and over the ensuing few million years, these combined with one another to create increasingly more complex molecular chemistry.** Chemical complexity now began to increase exponentially on a myriad of planets in the early universe.

Around **2.1 billion years after the Big Bang, complex carbon-based chemistry occurred on planets with suitable conditions. This evolved into DNA and other organic molecules across the universe, allowing the building blocks of life to form.**

Between **3-4 billion years,** (based on life on Earth), **the earliest life forms (single cell organisms) most likely emerged on innumerable planetary bodies where suitable conditions existed.**

At around **5-6 billion years, the first multi-cellular organisms presumably arose** (based on the evolution of life on Earth). It's possible that this may have occurred 2 billion years sooner on some worlds if the conditions were optimal.

Chemical complexity had increased to permit the first intelligent life forms, equivalent to present-day humans, to evolve. This may have taken half, to twice, as long as it did on Earth, depending on the conditions available. These timeframes assume that life is a natural consequence of the evolution of the universe. There is mounting evidence to support this.

After the earliest solar systems formed, **it took around another 7.6 billion years for our solar system to form and another one billion years after that for the first cells to evolve on Earth. It took another 3.8 billion years for humans to evolve and ultimately develop technology to take us to the Moon and send robotic spacecraft to other worlds.** Assuming that humans are not the most intelligent life form in the universe – a safe assumption – there may have been huge evolutionary leaps that occurred by other intelligences that allowed them to advance far further than we have billions of years ago before life on Earth even started to evolve.

The universe will not stop evolving at the level of humans because it surely has the potential to go almost infinitely far beyond the primitive stage of intelligence that we have reached. Many sciences are now revealing that **the universe has evolved towards increasing complexity.** To do this, the universe has to increase its information content. In order to do that, it must increase order and decrease disorder (entropy). By increasing order, this improves the chances of the universe continuing to evolve. Increasing disorder leads to its death, whereas order leads to stars, planets, life, intelligence, and increasing consciousness - and whatever evolves after that. Our universe may be the equivalent of the yoke of an egg. It may exist to allow consciousness to emerge and whatever greater existence that may lead to.

Some physicists are now starting to consider that the universe may be the ultimate organism. The human body consists of between 20 - 70 trillion individual life forms called cells. Could the universe be an entity whose 'cells' are super-intelligences?

History of the Universe

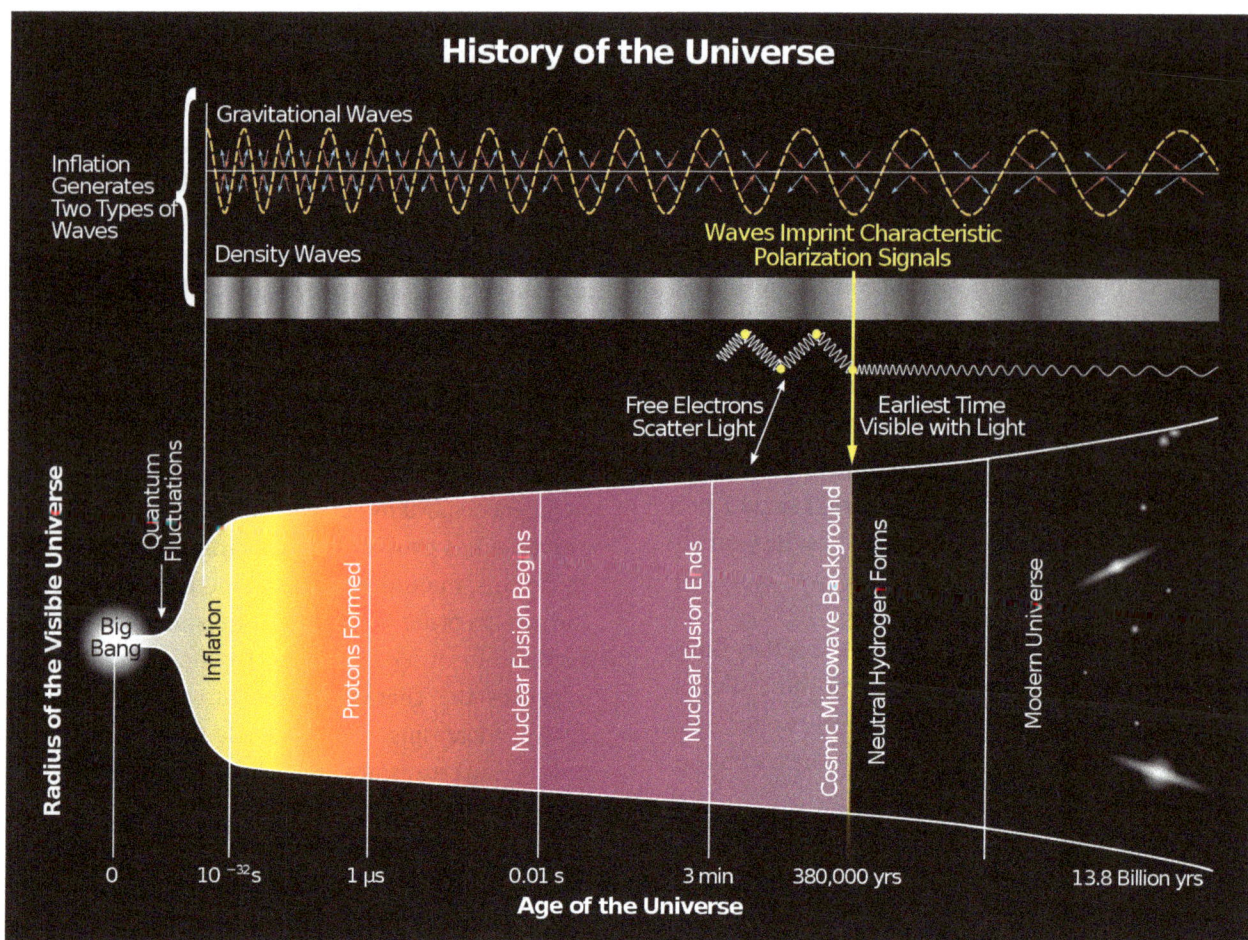

This diagram defines important stages in the evolution of the very early quantum universe.

CHAPTER 4

THE 'OBSERVABLE' UNIVERSE

The term 'observable' universe is often mentioned in astronomy, so it's important to understand what this term means. **The observable universe is the portion of the Greater Universe that we can observe**. This is similar to the amount of ocean we can see from a ship at sea. The observable universe is expected to be a very small part of the entire Greater Universe. The observable universe's horizon has a radius of around 13 billion light years from any point in it. We can see nothing beyond the distance because when we look deep into the universe, we see back to the time boundary when the universe formed.

The distance to the horizon on Earth is limited by our planet's curvature. The distance to the universe's horizon is limited by the speed of light. If the speed of light was infinite, then we would see the entire Greater Universe that lies beyond our observable universe. But because light travels very slowly, with respect to the size of the universe, this means that the deeper we look into space, the longer it takes light from distant galaxies to reach us. This also means that the deeper we look, the further back in time we are looking. **When we look back as far as we can, we see back to the birth the universe, where time began - so we cannot see beyond that.** This is therefore the boundary to the 'observable' universe. Nowhere do we see the universe as it is today. **Everything we see is sometime in the past, even our nearby the Moon and the planets.**

Our universe's horizon is both a distance horizon, and a time horizon. This stops us from ever seeing, or being able to interact with, the Greater Universe that lies beyond it. It's as if it does not exist. Galaxies that are beyond our horizon may as well be in another universe because no matter what happens in space-time beyond our observable universe, it will never affect us. This is because the universe is expanding faster than the speed of light. Information from the Greater Universe would have to travel faster than light to reach us, and that is not possible.

When we gaze across the ocean, we see a distant horizon beyond which we cannot see, but we know that the Earth extends far beyond the horizon. Similarly, astronomers know that our universe extends far beyond our horizon of our observable universe, possibly for an unimaginable distance. We know this because when the universe first formed, there were far more galaxies within our observable universe than there are today. This is because the universe was much more compact then, and the rate of its expansion was less. As the universe continued to expand, galaxies at the edge of our observable universe passed beyond our horizon. Given enough time, our galaxy will be the only galaxy within our observable universe. All other galaxies will have moved beyond our horizon.

If the universe had a constant rate of expansion, we would expect galaxies that are near the edge of the visible universe to have moved an extra 13 billion light years further away since the universe formed. But they have actually receded to around 45 billion light years due to the universe's expansion having increased over the last several billion years. However, the increasing rate of the universe's expansion is now thought to be suspect We can never see galaxies beyond 13 billion light years because they are receding faster than the speed of light, so their light can never reach us. *Space can expand faster than light but light cannot increase its speed, because it is finite and fixed.*

Always remember that when we look deep into the universe, we are not seeing galaxies as they are now, we are seeing them as they were when their light left them millions or billions of years ago.

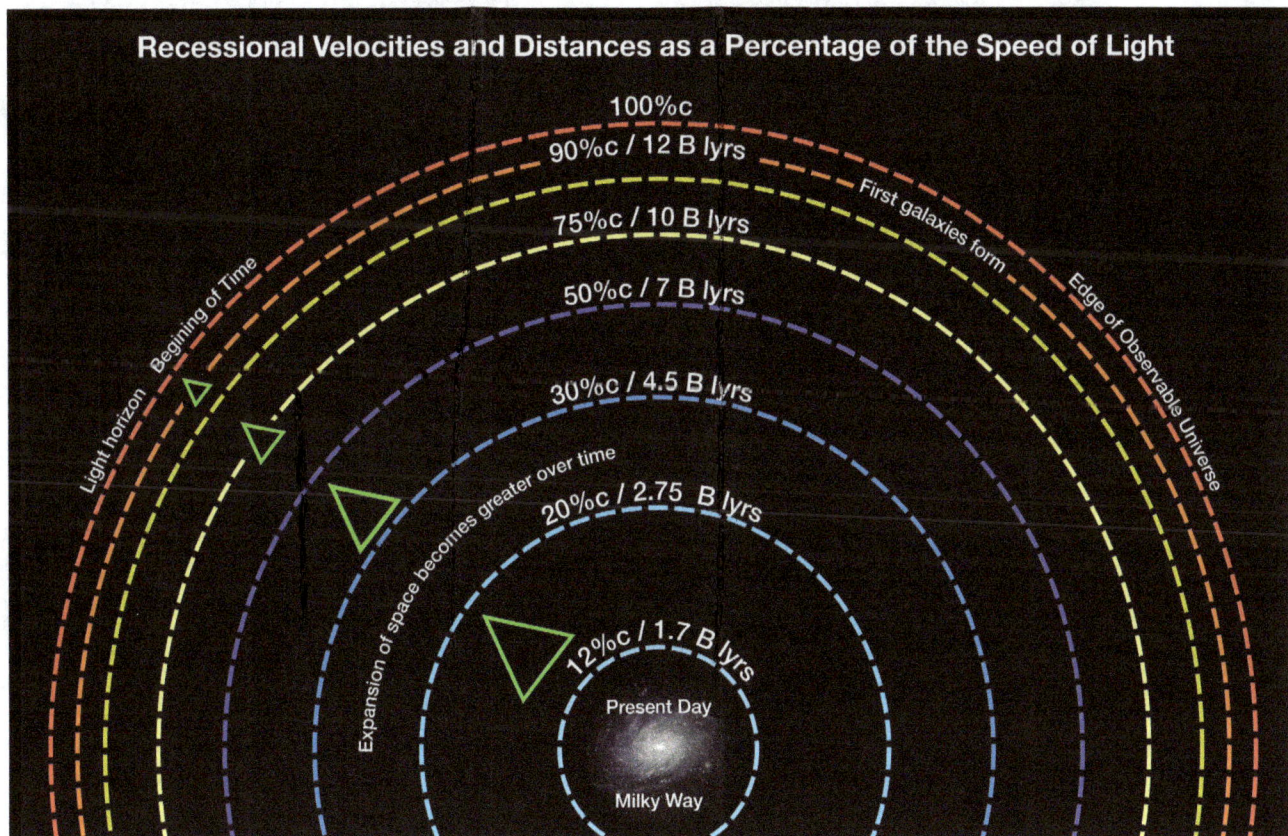

The distance to the edge of the visible observable universe is 13 billion light years. Note that the dimensions of space are represented as triangles. They become more contracted the further back in time we look. This is due to the universe being much smaller in the distant past than it is now after its expansion. The dashed circles represent the percentage of the speed of light (c) at which the universe is receding from us at those distances. The closest circle is 1.7 billion light years away Galaxies here are receding at 20% of the speed of light. The most distant one is 12 billion light years away. It is receding at 90% of the speed of light.

For astronomers living more than a century ago, the amazing concepts in this section would have been impossible for them to conceive. These concepts still remain inconceivable to most people alive today! So, you are one of a very small percentage of humans who understand this.

It took an exposure of 11 days using the Hubble Space Telescope's eXtreme Deep Field camera to capture this amazing image. It is of an extremely small part of the sky in Fornax that has an area equivalent to holding a 20 mm coin (a dime) at a distance of 23 meters (75') away! (Test how small that is.) There is only two stars with small spikes in this field. Everything else is a galaxy. There are approximately 10,000 or more galaxies in the original high-resolution image.

Galaxies become increasingly fainter, smaller, and more dense in number as they approach the edge of the observable universe due to their immense distance from us. At that distance, we see galaxies, as they were when they had just formed. Back then, galaxies were close together, and quite irregular. There are no galaxies visible beyond this boundary of the observable universe because they did not exist then. In this picture, young galaxies are blue due to them having masses of newly-born, very hot stars. As galaxies age, their stars consume their hydrogen gas so they become more yellow and orange. Old stars are only visible in relatively old galaxies. The colors of the galaxies tell us the age of most of their stars.

Obler asked. *"If the universe is infinite, galaxies should fill every part of the background sky, so space should be as brilliant as the stars that galaxies are made of. So why is the sky black?"* This was known as Obler's paradox. But for the sky to be as bright as the surface of stars, light would have to travel at an infinite speed so that there was no boundary to how far we can see. But light was measured to have a finite speed, so we cannot see an infinite universe. This limits our view of the cosmos to the observable universe which is just over 12 billion light years in radius. At that distance, there are not enough galaxies to over the whole sky to make it bright, so it is black. And due to light's finite speed, this also means that the further galaxies are away from us, the younger they look. This is due to the time it takes for their light to reach us. This means that at the edge of the observable universe, we were seeing back to the birth of the universe, so we cannot see beyond that because the universe did not exist before then. Credit: HST, NASA

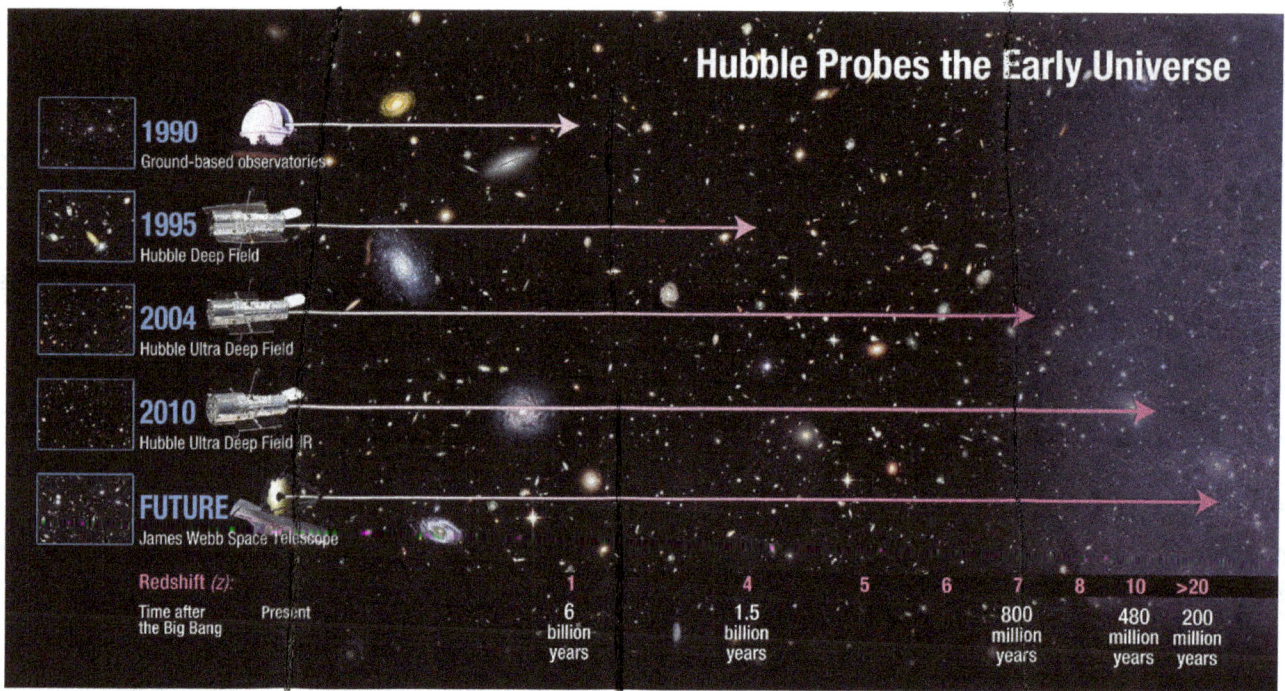

Hubble Probes the Early Universe

Redshift (z):			1	4	5	6	7	8	10	>20
Time after the Big Bang	Present		6 billion years	1.5 billion years			800 million years		480 million years	200 million years

Ever-improving technology allows us to look ever deeper into space, and therefore further back in time to see the early universe. Credit: NASA

CHAPTER 5

THE EXPANSION OF THE UNIVERSE AND WHERE DID IT COME FROM?

It's very important to reiterate that the Big Bang was an explosion of space; not an explosion occurring in some pre-existing space. Contrary to what illustrations and computer-art videos depict, there was no central point from which the universe expanded. Space in the Greater Universe expanded everywhere at the same rate. Because this is impossible to illustrate, the old myth of the expansion starting from one point will continue to remain a misleading part of how the Big Bang is visualized.

The science of the Big Bang does not say where space, energy, and time came from, or what made the universe come into being. It is often assumed that there was nothing before the universe was born, i.e. there was no space, and that time did not exist. However, the Big Bang theory does not make a claim one way or the other about how the Big Bang came about. It assumes that space, energy and time may have already existed, perhaps in another form as a single compressed force. It suggests that the universe was highly compressed in a singularity until something caused the Big Bang to occur. *If* this is so, it can be thought of like a spring that is wound so tightly that it becomes imperceptibly small. Then something happens that allows it to unwind at an incredible speed due to it having infinitely compressed energy.

But if space was infinitely compressed, then the three dimensions of space would be compressed into a one-dimensional point. Matter would not exist because it would be compressed into pure energy with no dimensions. Because energy is equivalent to mass, this infinitely small point should contain infinite mass and infinite energy. That being the case, the dimension of time would be so compressed that time would not 'tick'. So what could have made space expand, when there is no time in which it could expand? **How the Big Bang occurred is mankind's biggest mystery.**

I suspect that because we know so little about time and quantum physics, and almost nothing about the conditions created by infinite compression, we cannot see what might have caused the Big Bang to occur. Our present view is far too limited due to insufficient knowledge to visualize what the universe was really like at the very outset. In time, with the help of mega-computers, we will learn what did occur and how.

Space is able to *expand* far faster than the speed of light, but matter is not able to *move* faster than the speed of light. Space can expand or contract at any speed, but light (or anything else) can never exceed the universal speed limit of 300,000 km/sec in a vacuum, i.e. the speed of light. Why this particular speed is the limit for the speed of light is a major enigma for science to solve.

At the instant of the Big Bang, it is thought that the density of matter and the gravitational field that matter creates, must have been infinite (or close to it). All the matter that is in our observable universe, *and that in the Greater Universe beyond*, was compressed into an almost infinitely small volume. This would have made gravity, pressure, and temperature infinite. What ramifications this has is totally unknown.

Some of the conditions for the Big Bang appear to be the same as those for a singularity in a black hole.

(See Volume 3 Chapter 2, page 106.) However, there are differences. The Big Bang caused space to expand from what was a singularity (a one-dimensional point), whereas the infinite gravity of a black hole does the opposite: it causes space to *contract* into a singularity. If the universe was initially a singularity like a black hole is thought to be, then how did it manage to explode? Black holes cannot explode so what was different about the Big Bang? For one, black holes exist in pre-existing space, whereas, with the Big Bang there was no pre-existing space - that we know of. What makes the Big Bang and black holes so different should be something that cosmologists should try to solve because it may open new doorways into other aspects of cosmology and quantum physics. Some physicists have come up with some possible reasons to explain this, but they are long shots that are too complex and difficult to explain here. These concepts can be found on the Internet for those who are interested.

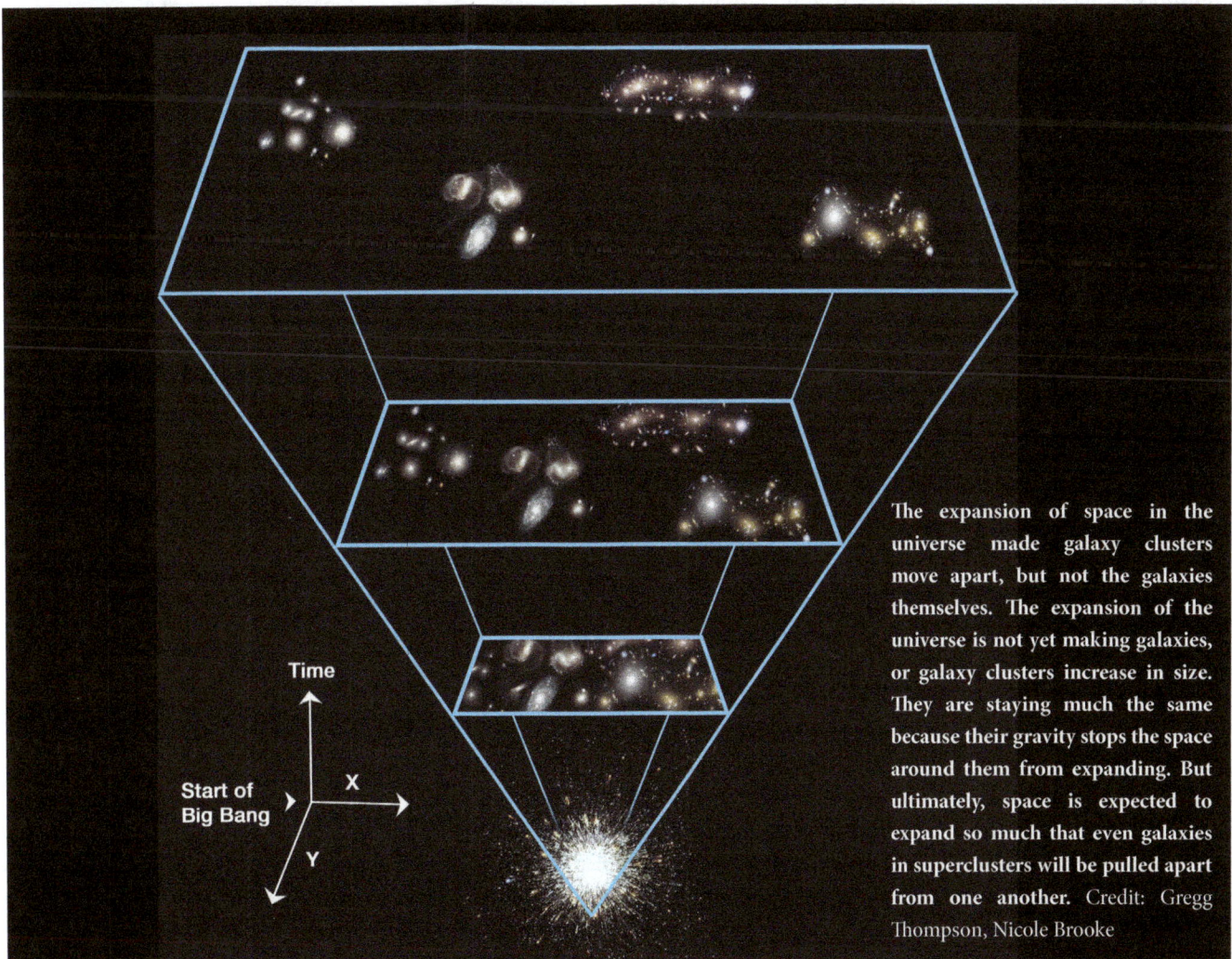

Time

Start of Big Bang

X

Y

The expansion of space in the universe made galaxy clusters move apart, but not the galaxies themselves. The expansion of the universe is not yet making galaxies, or galaxy clusters increase in size. They are staying much the same because their gravity stops the space around them from expanding. But ultimately, space is expected to expand so much that even galaxies in superclusters will be pulled apart from one another. Credit: Gregg Thompson, Nicole Brooke

WHERE COULD THE UNIVERSE HAVE COME FROM?

The following ideas for where the universe came from are highly speculative but worth considering.

In 1913, **Albert Einstein and Otto Stern calculated that even a total vacuum contains an energy field.** This field is called **zero-point energy**, or **quantum vacuum energy**. It is the lowest possible energy that a quantum physical system can have. It exists everywhere throughout the universe, and it has a high level of *potential* energy. All particles such as photons, atoms, and molecules oscillate rapidly giving them a high zero-point energy value. Even at just a fraction of a degree above absolute zero particles vibrate. The faster a particle vibrates, the higher its zero-point energy value becomes. Electromagnetic waves also have zero-point energy. This energy is the universe's ground state energy field. Einstein showed that energy and mass are equivalent. **Because the vacuum of space has a small amount of energy, it must also have a small amount of mass.** Discovering the exact properties of space is very important in understanding where the universe came from.

Quantum theory states that **a vacuum is seething with *virtual* particles**. It has been proposed that virtual particles may pop out of the vacuum field for extremely short periods before returning to it. Some cosmologists have speculated that an unexplained energy surge in a vacuum's ground-state energy field (that may have existed before the Big Bang) may have caused a virtual particle (a minute bubble of energy) to pop out of the background energy field (for some unknown reason), and not return to it. They propose that this 'energy bubble/particle' may have somehow expanded into the infinite energy field of the Big Bang. But they do not explain how! This is a very big long shot but explanations have to start somewhere.

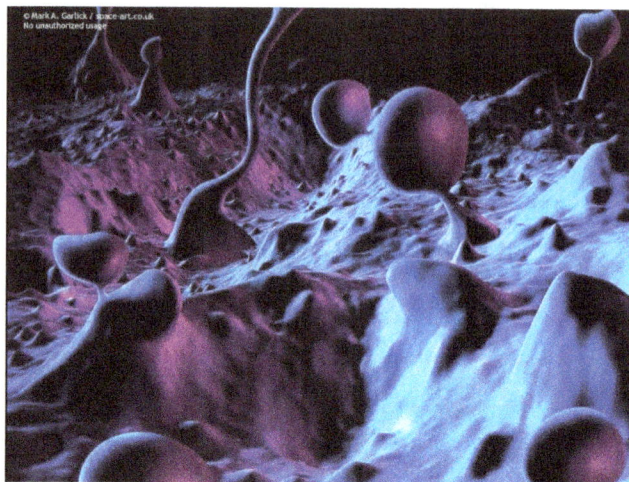

The universe's background energy field may appear something like this if it were possible to view it. Illustration by Mark A. Garlick

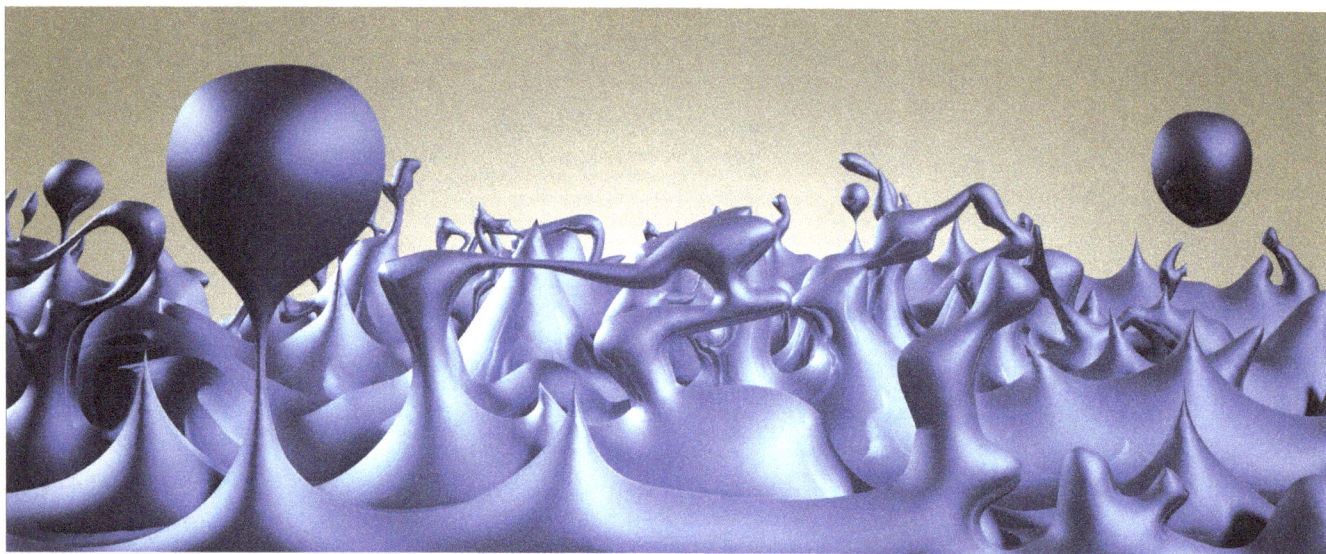

'Nothingness' (i.e. a vacuum), is thought to contain zero-point energy. The vacuum energy field (also called zero point energy) can be likened to a broth of virtual particles that act like water just before it boils. As water comes to a boil, bubbles and molecules pop out of the surface. Quantum theory suggests that virtual particles may be popping out of this electromagnetic, quantum 'foam' field for incredibly short timeframes before returning back to it. Virtual particles are calculated to contribute 90% of the energy in particles, such as a proton! Physicists speculate that an unexplained energy burst in the vacuum energy field might have caused a virtual particle to pop out of it and not return to it. For reasons not given, this particle may have multiplied exponentially fast and created the Big Bang. But How? Such speculation is fine, but it is nothing more until some solid science is found to support it. Credit: Robbert Dijkgraaf

It has also been suggested that the zero-point vacuum energy field may have kept randomly creating universes that were failures. Eventually, after a near infinite amount of time, one universe with exactly the right properties to evolve endless complexity emerged – ours! Such postulations require an explanation for where the zero-point vacuum energy field came from in the first place. Also, how did our universe come to be so incredibly finely-tuned far beyond random chance? (See Chapter 14.)

Cosmologist **Lawrence Krauss** delivers a lecture on YouTube titled '*The Universe from Nothing*' where he offers an explanation of how he believes gravity can permit the possibility of the creation of the universe, however, his explanation is another very long stretch, like all other explanations are.

Leading cosmologist **Alan Guth** believes that our greater universe may have come from another universe where dark energy caused space to contract to a singularity. Somehow, this may have caused our Big Bang to occur in a new 'space-time' which became our universe. This is a very long stretch. (See 'Dark Energy' on the following page.) **Lawrence Krauss** likes this idea despite there being no evidence for it and nor can there be because there is no way of knowing anything beyond our universe. If this could happen, one has to ask, "*Where did the original universe and its space-time come from*"? And "*What 'space' would our universe emerge into to start growing*"? The concept of universes being born from parent universes, in the same way that life reproduces is a cute idea. But it is passing the buck to another universe because we then have to ask, "Where did the parent universe come from?" Such explanations have as much credibility as believing that some god created the universe, a concept that Lawrence Krauss abhors!

All these speculative ideas are in the fantasy category at this stage, and it's unlikely that they could ever be tested. Cosmologists need to come up with plausible postulations to explain where the universe's energy came from. And these need to be testable against observation. Without this, we will never have a complete theory for the *origin* of the universe.

Other physicists are finding startling evidence that supports the possibility that our universe may be a simulation. Just as worlds in a virtual reality game are controlled by a computer program that exists outside the virtual world, so might there be some control program that exists outside the dimensions of our universe. If this were the case, then the energy could come from outside our universe, but this will be very hard, if not impossible to prove – unless we find the power cable inlet! But even if this were the case, we would then be back to asking, "*Where did the outside universe come from*"? I suspect the answer to this question is so far beyond a human mind. It is like expecting an ant to answer quandaries in physics when most humans can't.

CHAPTER 6

DARK MATTER &DARK ENERGY

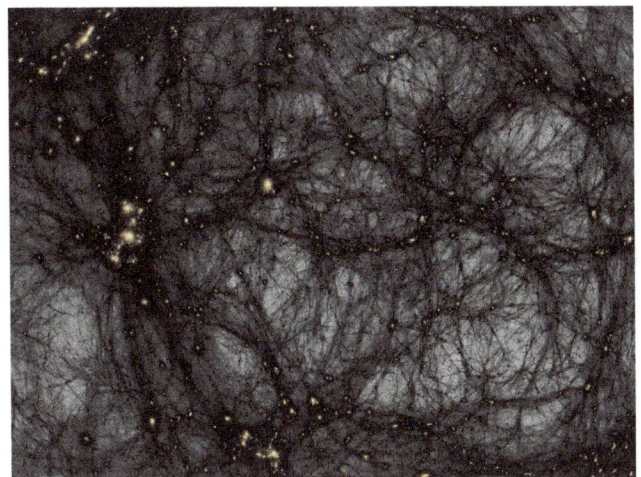

Dark energy appears to be what is repelling galaxies from the voids amongst the filamentary structure of the universe.

Dark matter is envisaged as an invisible, transparent material that does not emit or absorb any light. It surrounds galaxies (yellow spots) revealing its presence by interacting gravitationally with visible matter. Dark energy is said to act like an anti-gravity force. It pushes normal matter and dark matter away forming bubble-like voids between the filaments. This image is from the IMAX movie 'Dark Universe'. Credit: Kavli Institute of Particle Astrophysics & Cosmology at Stanford University and SLAC National Accelerator Laboratory.

The visible matter that we see in the form of stars, gas and dust in galaxies, *appears* to constitute just a mere 5% of the total mass of the universe! Whereas dark matter is *thought* to constitute around 26% of the universe's mass. But no one has determined what dark matter is! It is associated with bright massive stars, galaxies, and particularly with massive galaxy clusters. It acts like normal matter collecting together over time.

It now appears that *dark energy* must constitute the other 69% balance of the mass of the universe. Dark energy *may* turn out to be the fifth force of nature.

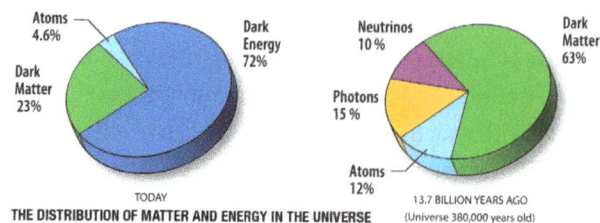

THE DISTRIBUTION OF MATTER AND ENERGY IN THE UNIVERSE

IMPORTANT FACTS ABOUT MASS, ENERGY AND GRAVITY

Mass is the most concentrated form of energy. One kilogram of matter contains 9×10^{16} joules of energy. If all that energy could be released, it would be enough to lift the entire population of the Earth into space! By comparison, nuclear energy releases only 1% of all the energy contained in matter. As matter disintegrates as it falls into a black hole, it releases 43% of its energy. This is why black holes emit huge amounts of energy in all wavelengths. Only a black hole has the ability to cause such a huge release of energy. When matter disintegrates upon entering its event horizon (the black outer edge of the black hole where time stops and gravity is infinite). When matter and antimatter collide 100% of their energy is also released.

Einstein showed that all forms of energy have effective mass and that they are therefore sources of gravity. Because the mass of a star curves space, planets are actually in free fall around their star. They never fall into it because there is no friction in space to slow them, so they endlessly go around it in their orbit trying to fall towards their star. It is the same with astronauts in the space station: they are in free fall around the Earth, so they are weightless. Before Einstein came along and discovered that mass curves space, it was thought that gravity was a force like magnetism that acted on all masses to draw them together. But Einstein showed that mass warped space to cause hollows in the space-time continuum. Smaller masses would 'fall' or 'roll' into the hollows that larger masses created. We misconstrued the curvature of space as being a force, which we called gravity.

Gravity (or more correctly, the curvature of space) is the organizing force for the cosmos. It brings matter together thereby creating structure to occur in the form of stars, planets, galaxies, and clusters of galaxies.. It is a very weak 'force' being a factor of 1036 or 1,000,000,000,000,000,000,000,000,000,000,000,000 times less than the fundamental forces which govern the sub-atomic quantum universe. But it is a cumulative force which acts on all matter in the universe over enormous distances. The other forces do not. The Strong and the Weak forces only act at incredibly small distances within atoms. Magnetism only extends around planets and black holes. Newton assumed that gravity acted instantaneously (infinitely fast), but when Einstein came along, he showed that it travels at the speed of light.

DARK MATTER

It is now realized that ordinary visible matter that constitutes nebulas, stars and planets in galaxies, is insufficient to prevent galaxies from flying apart as they rotate. Ordinary matter is only around 20% of the matter that is required to hold galaxies together. Where is the rest of the matter? To explain this, it was initially postulated that the extra matter required to hold our galaxy together might be normal matter in the form of trillions of small, dim red and brown dwarf stars that are too faint to detect, unless they are nearby. It was also speculated that the mass of gas in the huge halos that surround galaxies might explain some of the missing mass. (See Volume 3 Chapter 8.) Galactic haloes might contain numerous stellar-mass black holes, rogue stars, and large numbers of planetary bodies flung out of solar systems during close encounters with large planets during the formation of solar systems. (See Volume 2 Chapter 6.) But there is nowhere near enough observational evidence for any of these possibilities to constitute 5 times the amount of visible matter that is required to hold galaxies together. It was then thought that weakly interacting massive sub-atomic particles (WIMPs) were a possible explanation, but again, there was no evidence for these either. Some physicists then suggested that the dark matter mystery might be solved by modifying the theory of gravity so that on a large scale, gravity acts differently to how it acts on a small scale. But again, there is no evidence to support this, so dark matter remains the only explanation.. Cosmologists just don't know what dark matter is or how to test for it. They believe it's there, but it is a complete mystery as to what form it takes - if it is real and the cause.

It is thought that dark matter (illustrated here as the light blue haze), must be involved in galaxy clusters to hold the galaxies together. The denser the cluster, the more dark matter there needs to be. Illustration by NASA ESA JPL Caltech Yale CNRS

DARK ENERGY

In earlier decades, cosmologists assumed that gravity would gradually slow down the universe's expansion. In 1998, two teams of astronomers working independently of one another and without knowledge of the other decided to calculate the rate of the universe's deceleration by observing distant supernovas that occurred many billions of years ago. Much to their utter amazement, both teams found that **the universe's expansion was slower in the past than it is now. Around 4 to 5 billion years ago, it started to *accelerate*!** Why then? Both teams assumed that their results must have been due to observational error or miscalculation, so they were reluctant to publish their results. They did not know that the other team had come up with the same result. But after exhaustive checking, neither team could find an error in their work so they both published their findings. To their surprise, they found that another team had made the same findings that supported their observations and conclusions.

This meant that there had to be a repulsive anti-gravity force that was pushing matter apart to make the universe expand, but they could not find a source for this. For the lack of a better name, they called this force **'dark energy'**. Cosmologists calculated how much **dark energy was needed to cause the observed degree of acceleration in the expansion of space. To their surprise, they found that dark energy was the strongest force in the universe, having almost twice as much energy, and therefore mass as visible matter and dark matter put together!** This was yet another discovery beyond all expectations! The past has taught us to be wary of such extreme explanations when the evidence is flimsy.

Calculations showed that **as the universe expands, dark energy is not diluted: remarkably, it must retain the same force per cubic centimeter!** One would expect this to be impossible! How this could occur is baffling. We must remember that it is still early days, so there may be further unexpected discoveries that change our current assumptions made about dark matter and dark energy.

When **Albert Einstein** developed his theory of relativity, his equations predicted a repulsive gravitational force, however, he discarded this because he thought it could not be true. As stated earlier, this was because he believed, as nearly all astronomers did back then, that the universe was static and unchanging. To get rid of this anti-gravity force, he used a cosmological constant in his equations. When Hubble discovered that the universe was indeed expanding, Einstein regretted that he did not trust his calculations. Einstein's equations said that the vacuum of empty space must contain vacuum energy. Some physicists think this is what dark energy is. If so, empty space could be what is making the universe expand. But why is the expansion getting faster? And why is it strongest in the voids between galaxy clusters?

Dark matter attracts normal matter while dark energy repels both forms of matter. Dark energy explains why space is stretching and causing the universe to expand. Given that space and time are interwoven as the space-time continuum, this means that dark energy is not only stretching space, but it should also be stretching the passing of time, causing it to slow down. (Refer to page 37.) How could this be tested? If time is slowing down, we would not know. This would make the expansion become correspondingly slower, but to us it would appear faster.

All this is based on the assumption that the universe is expanding ever faster? But is it? A team of astronomers led by Perlmutter, Schmidt, and Riess made observations of distant exploding stars known as Type 1a supernovas that explode in a very regular way compared to other supernovas. By measuring their brightness, they could determine how far away they were, because the fainter they are, the further away they are. And the further away a supernova explosion is, the more its light is red-shifted. This is due to the expansion of the universe causing the Doppler effect, which stretches a star's light into the red end of its spectrum. And the further away a supernova is, the further back in time we are observing it. The brightness and the redshift of a supernova is directly related to its distance from us. But the Perlmutter team found a difference between these results that they could not explain, so they went and checked all their results for an error, but there were none. They interpreted this as being due to the expansion of the universe increasing exponentially around 6-7 billion years ago. They put this down to dark energy becoming increasingly stronger. In 2011, they won a Noble Prize for their observations. However, their explanation is now in doubt.

A 2019 paper by astrophysicists Colin, Mohayaee, Rameez, and Darkar, explains that Perlmutter's team's observations were made well over 20 years ago and that they used a relatively small sample of only 110 supernovas, which were mostly from one direction. They did not have thousands from across the whole sky, as many assumed. Since then, data has been collected from over a thousand Type 1a supernovas from across the whole sky. This included the rate at which they were moving away from our galaxy – data that Perlmutter's team did not have back then. When all this data was analyzed, this produced no major difference between their redshifts of supernovas and their distance. Nor did Perlmutter's team think to take into account that our galaxy is moving through space in one direction at a particular rate, while many others are moving away in different directions at different velocities. Light from supernovas that are traveling away from us at high speed are more redshifted than those traveling more slowly or at right angles to us. When this is taken into account, the increasing expansion of the universe disappears.

David Wiltshire and many other astronomers are now starting to think that dark energy may not exist.

There are far too many things that do not make sense. But once we know enough, we will find that they do make sense. Science has been in this situation innumerable times and it has managed to find answers to most dilemmas.

CHAPTER 7

THE MULTIVERSE CONCEPT

In my mid-teens when I had learned about the basic structure of the universe, I conceived the idea that our universe could be nothing more than an atomic particle in a much larger universe than ours. And possibly, this would be no more than a speck in an even larger universe, and so on. Some decades later, I discovered that cosmologists and others had also come up with much the same idea. The reason for this might be that the more we learn, the larger and more complex our perception of our 'universe' becomes. When we look into a microscope or a telescope, we see seemingly endless levels of detail with one level being a small part of the one before it. So, it's reasonable to extrapolate this into their being one cosmos inside another and so on – like Russian dolls.

Throughout human history, our perception of our 'universe' has changed remarkably. For cave dwellers, their universe was the land that they roamed across to as far as they could see. To them, the sky was like a mysterious ceiling. Hundreds of millennia later, when people traveled much further afield, the known area of land and sea became much larger. This planet became their 'universe'. Later on, some Asian cultures proposed that our flat Earth floated in an endless, fluid-like space. But around 2,300 years ago, **Eratosthenes** calculated that the Earth had to be a sphere. Incredibly, it took until the 1500s for this concept to be accepted! The Earth and the other planets were then known to orbit the Sun in our solar system. With this knowledge, the universe had become much larger than the Earth. But it was not known then what stars were. In the 1900s, it was confirmed that our Sun was just one star in a huge body of other very distant stars. The universe now became much, much larger than our solar system. But, what we thought was our universe then turned out to be just a part of our Milky Way. It was then discovered that we live in a huge spiral galaxy filled with a trillion stars! But once again, our universe grew enormously larger when we realized that the multitude of faint hazy objects in space were not gas clouds in our galaxy condensing into solar systems, but rather, they were other distant galaxies like our Milky Way! This one discovery made our universe become billions of times larger! Today, we now realize that our galaxy is but a speck in a universe filled with trillions of galaxies that stretch across space to the edge of the observable universe and far beyond.

Since the beginning of the rise of science, the more we learn, the more our universe grows ever larger and more complex. Will this continue?

Some cosmologists are now wondering if our universe might exist in a super-space full of other universes. This concept is referred to as the 'multiverse'. Some speculate that the multiverse may be giving birth to ever more universes all the time. However popular this concept has become, there is no evidence whatsoever for a multiverse, and nor is there ever likely to be because it is beyond our universe, so we can know nothing about it. The multiverse is the current new craze amongst cosmologists who should know better than to get on this bandwagon because their colleagues are on it, and because it excites those in the public that want a new form of heaven, whereby they will not die.

The Hindu culture thought that what little they knew of our planet was supported by giant elephants that stood on the back of a giant turtle swimming through some universal ether. Our perceptions of a multiverse are most likely just as naive.

At the very end of the first *'Men in Black'* movie, the camera pulls back from the Earth to see it and the Moon disappear into our solar system, which then recedes into our Milky Way galaxy as the camera travels though innumerable stars until we leave our galaxy behind. Beyond this, we see numerous other galaxies in our universe. The camera continues to pull back and we see our universe as a tiny sphere amongst other spherical universes. As the camera pulls back further still, these universes are then seen as being inside marbles, which three-fingered, super-alien children of an incomprehensible size are playing with them! The audience is left wondering what sort of universe do these super-sized aliens exist in, and if there is no end to this progression of ever-larger universes. I'm sure many people across the world have thought of this concept because it is a natural progression from what we have learned to date.

Some cosmologists are now catching up with these simplistic concepts by inventing the multiverse, but I'm sure it is as naive as my childish ideas were of ever larger universes inside one another. If there is anything beyond our Greater Universe, then it will be far, far beyond anything that the most imaginative human could ever comprehend – just as our solar system and our galaxy were beyond the ability of primitive cultures to conceive of or comprehend.

As a multiverse can only exist outside our universe, there is no way to prove, or disprove it. In science, the inability to disprove a thesis is just as big a problem as not being able to prove it.

Some current concepts suggest that there is a multiverse in which a universe exists for every possibility. However, the rules of logic state that when there is a concept like a multiverse where it involves infinities and anything is possible, then one possibility must be that it is impossible – as is the case with String Theory.

Cosmologist **Alan Guth**, who conceived the concept of inflation of the universe, went on to find observational evidence that strongly supported his theory of inflation. He claims that his recent calculations show that half the dark energy in the universe becomes pockets of space-time that ultimately contract exponentially into singularities. (For singularities, see Volume 3 Chapter 2, page 105.) The other half of the dark energy causes space to expand faster than the regions that are decaying. But why? Guth uses this to ensure that the expansion of the universe will not end. To date, there is no evidence for or against this concept. It sounds like an each-way bet.

This illustration attempts to visualize a very small portion of the hypothetical multiverse. In it, there is an infinite number of bubble universes that breed infinite numbers of new bubble universes. These, in turn do the same thing ad-infinitum forever! Where would all the energy for this concept come from? And what would be the reason for a multiverse? When infinities are introduced, they typically mean there is an error.

Guth speculates that at high-energy states, such as during the Big Bang, or when some dark energy is compressed into a singularity, gravity can act *repulsively* making a singularity explode. However, Guth does not explain why this does not occur with singularities in black holes. Guth wonders whether our universe could have been created from a region of space in a parent universe that decayed into a singularity and then exploded to form a new universe in a new space-time. *If* this is possible in a mega-universe, then infinite numbers of embryonic universes would be being born into whatever super-universe our universe came from. If our universe could give birth to other universes, then those universes might do the same. Where would it end, if ever? There would be infinite numbers of universes to the power of infinity! This would be mathematically and physically impossible.

If there are other universes, String theorists suggest that each 'pocket' universe could be built around a different vacuum energy level. This would make each universe different. This is total speculation with nothing to support it. There is even dubious evidence to support Santa Claus.

Our universe required an extreme level of fine-tuning to exist, so the chance of other universes having such conditions by *pure chance* should be impossible. If other universes do not have the right conditions, they would be very short-lived, extremely simple, or stillborn. However, physical laws in other universes may be quite different so some may be viable. But why would there need to be more universes other than our Greater Universe?

If the multiverse is possible, one would have to ask, *"Was there an original super-universe from which the multiverse was born, and if so, where did that super-universe come from, and how was it created?"* A multiverse does not seem to get us anywhere. It just pushes the answer as to where our universe came from back one more step. Cosmologists come to this conclusion again and again.

The '13th Floor' movie speculated about what could happen when virtual worlds are invented that replicate real worlds. This is a scene from the movie where the main character drives away from his virtual city and through a barrier to find he had reached the edge of the city's simulation and discovered it is not real. Credit: Centropolis Entertainment

An analogy for the **multiverse concept** could be an infinite forest. Each tree produces large numbers of seeds. The seeds that are lucky enough to land in a rich growth medium grow into slightly different trees to their parent tree. Most do not germinate at all, and many of those that do, die soon afterward due to unsuitable initial conditions. New trees that survive produce multitudes of seeds before they die and the cycle repeats. But remember, all this is extreme speculation with not a skerrick of evidence to support it. And because all these concepts lie beyond our universe, it would be impossible to find any evidence to test whether any of them exist.

COULD THERE BE PARALLEL UNIVERSES?

Some cosmologists have their own personal form of a multiverse - just like most religious people have their own version of their religion: they keep the bits they like and can live with, and reject the rest. Some cosmologists think there is a multiverse that allows for every possibility that could ever exist – an infinity of so-called, parallel universes. This would be impossible because each parallel universe would require an infinite number of positions for every particle in every universe for every nanosecond throughout all of time! At every nanosecond, zillions of zillions of new 'parallel universe would be branching off from every universe into a new universe which would do the same because every single particle would have infinite possibilities. The number of possibilities in every universe would grow exponentially every nanosecond. This is incomprehensible. Such a concept amounts to infinity multiplied by infinity multiplied by infinity an infinite number of times!!! The parallel universe conjecture is the most preposterous and naïve of all concepts. It would be far worse than the simplistic examples believers in this idea offer, such as Elvis does not die early in one universe, but instead, an infinity of other universes diverge from every nanosecond of his life. In many of them, he would never die! There also has to be every conceivable possibility for when, how, and where he finally dies. It would also have to include many possibilities where he lives long enough for technology to keep him alive forever. The same would be true for every place he could ever go, for every person he could ever meet, every song he could ever sing, every experience he could ever have, and so on.

This would also be true for every person that has ever lived throughout all of time, and similarly for every life form, and every possible future of our planet, and every solar system, and every galaxy, and so on. This would not be limited to our universe, but to every other universe! Clearly, **this is insane at every level.** Parallel universes and the multiverse are concepts that are nothing more than feel-good ideas for some people who want more options in their lives. It amounts to sheer nonsense to the power of infinity!

If there could be a near-infinite number of universes, some cosmologists think that our extremely finely-tuned universe would have had to eventually come into being by pure chance. However, mathematicians have calculated that the chances of this occurring are actually zero, even given infinite time for this to occur. This is because our universe required an incredible degree of fine-tuning across so many of its parameters that this would be impossible to achieve by pure chance. It would be infinitely more unlikely than expecting an ape to type up 'War and Peace' by pure chance. Even if the ape was given infinite time, probability theory makes it clear that the ape would not even type up half a page by chance, let alone the entire book.

The multiverse concept has been around for many decades and there are a number of prominent physicists and cosmologists, who cautiously support this crazy idea. They all have their own ideas about the types of universes that might exist, and how the multiverse would work. I doubt that any of them have really thought through how infinitely impossible it would be. Without any evidence to support this concept, the multiverse concept is more akin to a religious belief or a philosophy than it is to being a scientific theory. A leading cosmologist, **Paul Steinhardt** sums it up by saying, "*A theory that predicts everything, predicts nothing.*"

VIRTUAL UNIVERSES

Humans are now creating increasingly more realistic 'virtual' or 'simulated' worlds in computer games. In the not-too-distant future, super intelligent computers that are being developed at present will have unim aginable knowledge, intelligence, and creative abilities. They should be able to create, incredibly realistic virtual worlds. Given that there are likely to be numerous other super-intelligences in the universe that have evolved over billions of years, they could have already done this. And in each virtual world, numerous intelligences could evolve in them as well. They might even create virtual galaxies, so there could be a near-infinite number of them also. But this would require virtually infinite energy to run them. Where would this energy come from? There seems to be an underlying flaw in this scenario as well, because, when infinities are introduced, it typically means there is a mistake that has been overlooked. This concept, like those before it, is in the realm of fantasy rather than reality.

Could force fields permeate higher dimensions in our universe, or possibly connect to other universes in ways that today's science cannot imagine?

Physicists from Oxford University believe they have shown that entire universes cannot be simulated by mega-intelligent computers. They discovered a link between gravitational anomalies in the underlying space-time geometry and computational complexity that make it impossible to construct a computer simulated universe due to a quantum phenomenon that occurs in metals that exhibit strong magnetic fields at very low temperatures. They found that the complexity of the simulation increases exponentially with the number of quantum particles being simulated. The amount of computing power required for a simulated universe would double with each particle added, so the task would quickly become impossible. They calculated that just storing a couple of hundred electrons would require a computer memory that would physically require more atoms than exist in the universe! Their work seems to provide convincing evidence that entire universes could not be simulated because the quantum universe would not allow it. This may also apply to whole planets let alone entire solar systems.

If it was possible to somehow create a universe, then there would be little, if any, difference between a virtual universe and a real one, so it wouldn't matter to us one way or the other which one we lived in. What would define a 'real' universe anyway? The concept of creating a small virtual world was the basis for the movie '*The 13th Floor*'. Worlds may be simulated, but not entire universes. This may also apply to whole planets let alone entire solar systems.

CHAPTER 8

THE PECULIAR PROPERTIES OF SPACE AND TIME

Einstein's genius was to realize that space and time are 'elastic' in that they both contract in a gravitational field. When the field reaches infinity, they contract to zero. Such is the case with a black hole. He also showed that this also occurs under acceleration where, at the speed of light, length in the direction of the acceleration contracts to zero, and time stops. Einstein showed that space and time are two sides of the same coin, which he called space-time. There is much observational evidence for this. (See Volume 2 Chapter 6, page 198 and Volume 3 Chapter 2, page 59 to see how gravity curves space-time.)

An electrical field and a magnetic field were found to be expressions of the same force. Similarly, Einstein also discovered that mass and energy are equivalent, and so are gravity and acceleration. Light and other particles act like solid particles under certain conditions, yet under other conditions, they act as if they are waves. We live in a universe of equivalents, but these equivalent properties appear quite different to us. Are each of these 'equivalents' manifestations of some underlying force seen from different perspectives? These equivalents are telling us something very important about how the universe is structured, but we have not yet grasped the significance of this.

We need to integrate the theory of quantum mechanics with the theory of relativity but to do so, we need to understand the true nature of the quantum universe. However, the complexity of this could be like expecting our Stone Age ancestors to have conceptualized that their world was not flat as it seemed, but huge and round. And on top of that, they would need to understand that everything is made of these incredibly small things we call atoms. And furthermore, their world is suspended in a medium we call space and it goes around the tiny Sun in the sky, which in turn, is orbiting the center of the milky stardust scattered across the night sky! No part of this concept would make any sense to Stone Age people. We may very well be as far away from knowing the true nature of the workings of the sub-atomic universe and how our universe came into being as Stone Age man would be about what we know.

THE PROFOUND EFFECTS OF RELATIVITY

The following aspects of physics are most curious.

There is no place in the universe that is not in motion, and there is no place that is not experiencing some gravitational attraction, so space and time are always contracted to some extent. Clusters of galaxies are accelerating towards one another due to the pull of each other's large gravitational fields. Because the expansion of the universe is expected to keep going and maybe accelerating, the effects of relativity should make space shrink to some degree in the direction in which it is accelerating. If the acceleration approaches the speed of light, then space would shrink to zero. Cosmologists do not appear to have considered this.

Einstein showed that there is no absolute frame of reference in the universe because there is no center to the universe that can be considered at rest. This means there is no way to measure an object's velocity in relation to a stationary point. The velocity of an object can only be measured in relation to other objects that are also in motion. To get around this, astronomers consider our galaxy to be at rest. Einstein said, *"Everything is relative to the observer"*.

The speed of light is always the same in the vacuum of space *no matter how fast the source of the light is traveling*. Inexplicably, a light beam could be emitted from a fictional spaceship traveling at 90% of the speed of light, yet the light beam will not travel at 190% the speed of light: it will always travel at the speed of light and never exceed it regardless of the speed of the source of the light. This seems impossible, but it is a well-proven fact.

Einstein also proved that time is relative to the observer. It ticks at a rate that is relative to the observer's acceleration and/or the strength of the gravitational field that the observer is in. So, if a person could travel at almost the speed of light, they will experience the same slowing of time as a person who is in a near infinite gravitational field like that of a black hole. Because space and time can stretch and contract like elastic, this keeps the speed of light the same for all observers regardless of their acceleration or the strength of the gravity field they are in.

Imagine a fictional person has spiraled into the inner accretion disc of a black hole. He is spinning around it at say, 99% of the speed of light. To him, his watch would appear to be ticking as usual, whereas observers outside the effects of the hole would see his watch almost stopped! As he spirals in towards the black hole's event horizon where time stops ticking altogether, he feels like he is barely moving, yet the observers outside the hole see him whizzing around it at nearly the speed of light. Due to the effects of relativity, when he looks outward, he sees everything in the outside universe moving incredibly fast. It is evolving so fast that he can see to the end of time! This occurs because, for him, time has almost stopped ticking due to him being in the black hole's infinite gravitational field, so for him he is there for a very long time. But to him, it seems like no time all! To him, his watch appears to be ticking as it was before he started spiraling into the black hole. These opposing occurrences happen, as Einstein proved, that everything is relative to an observer's location and therefore his conditions - meaning the speed at which someone (or an object) is traveling, and the strength of the gravitation field they are in.

THE PHOTON PARADOX

The speed of light creates the most curious paradox for photons.

Because we experience a four-dimensional universe (i.e. three physical dimensions and one of time) we observe photons (particles of light) traveling through space and through time. However, if we could sit on a photon, the universe would appear to be a dimensionless point where neither time nor space exist! To a photon, its entire universe is a point, known as the zero dimension. This dimension is what the universe would have been just before space and time were created during the universe's birth.

We don't know much about time other than the arrow of time always travels from the past to the future, but in the mathematics of physics it can go either way! It is associated with a transition from order to disorder e.g. from life to decay, or from a new building to a ruin. Over time, everything decays- even the universe.

As stated, Einstein proved that the faster an object goes, the more the passage of time slows until, at the speed of light, time stops ticking altogether. Because photons always travel at the speed of light,

the dimension of time does not exist for them. Therefore, **in the photon universe, nothing can change because there is no time in which change can occur.** With his Theory of Relativity, **Einstein proved that length shrinks to zero at the speed of light in the direction of motion. This means that there is also no distance in the photon universe** because, for photons, every part of the universe in their direction of motion would be at the same point, the zero dimension. Space has no length at the speed of light. And because distance does not exist for photons, this is another reason why time does not exist - *it takes no time to go no distance.* **Along a photon's path, a photon would seem like it was at all points in the universe at the same time.** If we could travel at the speed of light, it would take no time to go to what people in the below-light-speed, 4-dimensional universe call infinity!

In summary, photons exist in the zero dimension that has no spatial dimensions, and no time. Because we exist in 4 dimensions, we can see photons traveling across space and through time. This demonstrates what a difference extra dimensions make.

In considering the photon paradox, also remember that photons can appear like a particle while at other times, they act like a wave. They are most likely neither, but rather something far more exciting but which we are unable to imagine.

CHAPTER 9

THE GEOMETRY OF SPACE

Just as primitive people wondered how far their earthly 'universe' extended beyond their horizon, today astronomers are wondering how far space extends beyond our observable universe. At least primitive people could walk beyond their horizon because it was stationary. But because the universe is expanding faster than the speed of light, even if we could travel at near the speed of light, we would never get near the edge of our observable universe because it is moving away from us faster than the speed of light due to the expansion of the universe being faster than light. To know what the greater universe is like seems impossible because there is no way to observe anything beyond our observable universe. And because of this, no information can ever reach us from beyond the observable universe.

In the future, we might be able to observe galaxy clusters in great detail that are more than halfway to the boundary of our observable universe (6-7 billion light years away). In doing so, it may be possible to detect gravitational influences on them caused by supermassive galaxy clusters that are beyond our observable universe but within theirs. If we could observe such effects, we would see them occurring 6-7 billion years ago.

IS THE GREATER UNIVERSE THE SAME AS IT IS IN OUR REGION?

We assume that the greater universe beyond our observable universe looks the same as it does in our region, but does it? At least the nearby greater universe should look the same because it was originally inside our observable universe. It's possible that the greater universe could have many different environmental regions like the Earth does. It has oceans, river systems, ice caps, deserts of rocks and sand dunes, rainforests, grasslands, wetlands, plains and mountain ranges etc. Could there be environmental variations in the Greater Universe like we have on Earth?

Could the density of galaxies vary across the greater universe? Could gravity have different strengths in different locations, as it does on Earth from the bottom of the ocean to the top of the largest mountain ranges? Could time work differently in some places? Could space be more compressed or expanded in different directions? Does the universe expand evenly all over? Is it possible that the Greater Universe is rotating so that its velocity varies from one region to another? Could the geometry of the Greater Universe vary like the surface of the Earth does from deep ocean trenches to high mountains and flat plains?

We now know how large the observable universe is. As far as we can tell, the geometry of space in the observable universe appears to be flat. But that may be an illusion because the curvature of the Greater Universe could be so slight due to it being so large that we cannot detect any curvature – just as an ant would not detect the curvature of our Earth by walking across a salt lake a kilometer wide. If astronomers could determine the geometry of space, this might help us determine how large the Greater Universe might be. But how could that be done?

DISCOVERING THE GEOMETRY OF THE EARTH

In 240 BC, Eratosthenes brilliantly worked out the shape of the Earth and found that it was round. Incredibly, he did this by using two shadows and simple geometry.

For all of humankind's existence, the Earth was considered to be flat until around the 6th C BC when Greek philosophers thought it might be round. It took another 300 years before **Eratosthenes** came along and conceived a simple method of measuring the Earth's diameter and its circumference and in doing so he proved that the Earth's geometry was indeed a sphere.

Eratosthenes was a genius who lived in Egypt. He was a Greek astronomer, a geographer, and a mathematician. He made many major contributions to science. He came up with the system of longitude and latitude, a calendar that included leap years, and a system for identifying prime numbers. He also invented the armillary sphere to predict the motions of the stars across the sky. He accurately measured the tilt of the Earth's axis and he compiled a star catalog containing 700 bright stars. It appears from the records of others that he also may have made measurements to estimate the distance to the Sun and the Moon. If he did, disappointingly, his records of this have been lost. Without doubt, he was an extraordinary genius.

Eratosthenes knew that the Sun did not cast a shadow at midday on the day of the summer solstice in the town of Syene in southern Egypt, which is on the Tropic of Cancer. ISunlight shone down into the bottom of a deep well because the Sun was directly overhead. But at Alexandria, almost 840 km (530 mi) in a direct line due north of Syene where Eratosthenes lived, the Sun *did* cast a shadow at midday at the summer solstice. The fact that the shadow lengths were different made him think that this might be due to the Earth being a sphere, as others before him had suspected. He concluded that if it was a sphere, then he could calculate the Earth's diameter by geometry using the differences in the shadow lengths in both towns. He did this by measuring the length of the shadow cast by a tall tower in Alexandria at midday at the summer solstice. By knowing the height of the tower, he was easily able to calculate the angle of the shadow to the Sun. He used simple geometry taught in most schools today in grade 9 or lower classes.

He measured the angle of the shadow in Alexandra to be 7.2° from the vertical. This meant that the tower was leaning away from the vertical walls of the well in Syene by that angle. He cleverly thought that, if the Earth was a sphere, then a hypothetical line extended into the Earth from the well in Syene, and another one from the tower in Alexandria, would meet at the center of the Earth. (See the diagram following.) The angle of 7.2° happened to be 1/50th of a circle, so he knew from geometry that if he could accurately measure what the distance was between Syene and Alexandria, then all he had to do was multiply that distance by 50 and he would have the Earth's circumference! That was exceptionally clever for over 2,200 years ago.

Ⓒ For fun, you and a friend in another town could do this experiment by measuring, the angles of the shadows cast by vertical sticks in each town at midday on the same day. This will allow you to calculate the distance to the center of the Earth as long as the distance between the sticks is accurately known. GPS systems or Google Earth can be used to determine this distance accurately.

Back in Eratosthenes time, measuring the distance between Stene and Alexandra was a very difficult task. This was because there were no straight, level roads between these towns, and there were no accurate maps. So, Eratosthenes decided to measure the distance by hiring a man to count the number of steps he took between these two towns as he walked along the road beside the Nile River. The road was relatively straight much of the way and it did not have any mountains to go over, so Eratosthenes did not have to allow for the extra distances traveled in the 3rd dimension. He made an allowance as best he could for the curves in the road. This was nevertheless a very difficult task for both Eratosthenes and particularly the walker as he had to walk close to 1,000 km (600 mi) counting every step.

Eratosthenes used the length measurement of his day, a stadio, to make his estimate. But there were two different lengths applied to a stadio, so historians are not sure which one he used. Because of this, we do not know whether his estimates for the radius of the Earth and its circumference were accurate to an amazing 1% or, at worse to 16%. Either way, Eratosthenes was remarkably close to what we know the Earth's radius to be today. By doing this, he confirmed that the surface of the Earth was not flat but curved. This explained why ships would sail over the horizon and not fall off the Earth. It also explained why we can see far further from a mountain than we can from a flat plain. In honor of this remarkable genius, a large, interesting crater on the Moon is named after him. (See Volume 2 Chapter 5, page 156.) Such

brilliant, yet relatively simple thinking goes to show what the human mind can achieve. Could we do something similar with the observable universe to discover the geometry of the greater universe?

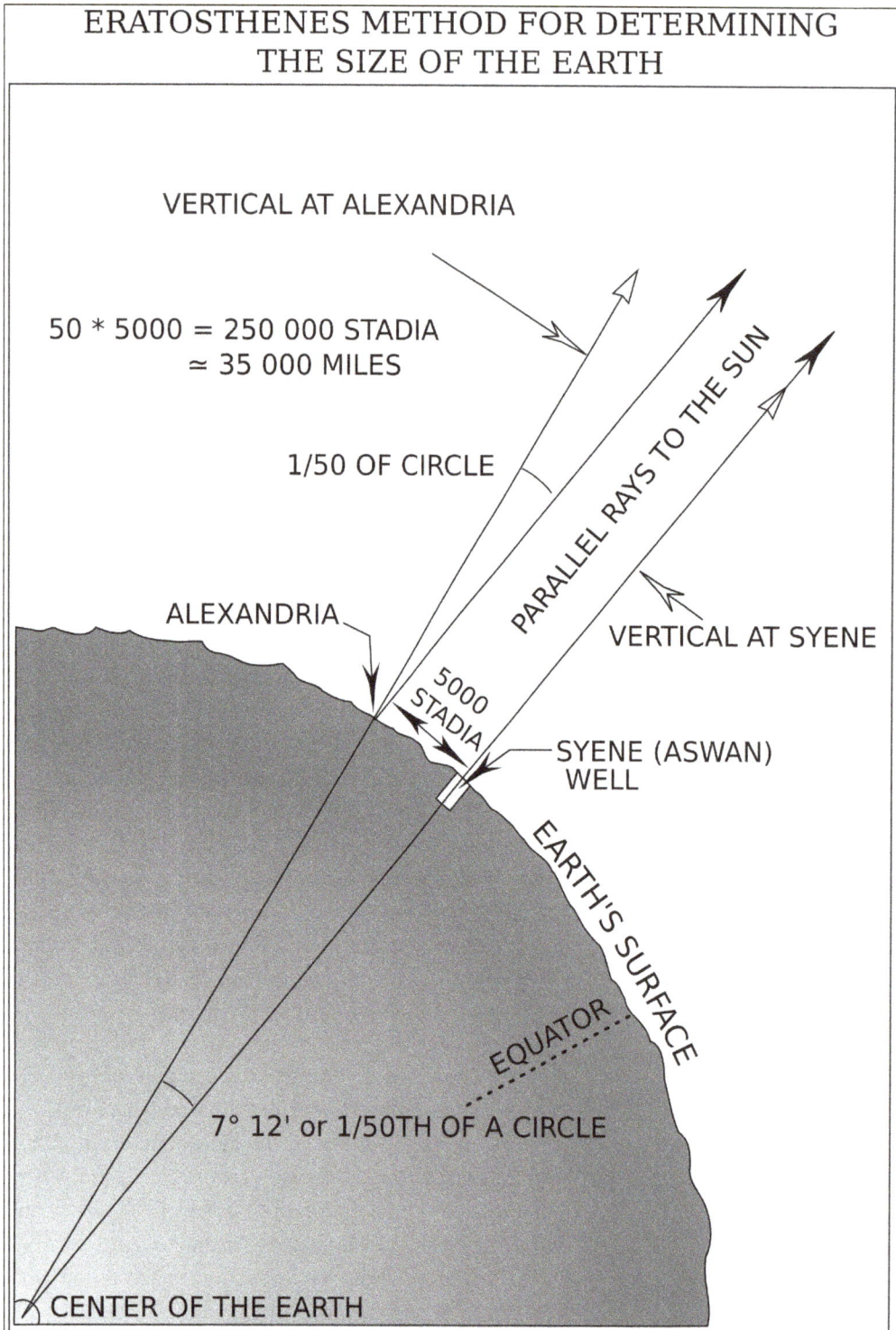

ERATOSTHENES METHOD FOR DETERMINING THE SIZE OF THE EARTH

VERTICAL AT ALEXANDRIA

50 * 5000 = 250 000 STADIA
≈ 35 000 MILES

1/50 OF CIRCLE

PARALLEL RAYS TO THE SUN

ALEXANDRIA

5000 STADIA

VERTICAL AT SYENE

SYENE (ASWAN) WELL

EARTH'S SURFACE

EQUATOR

7° 12' or 1/50TH OF A CIRCLE

CENTER OF THE EARTH

The geometry Eratosthenes used to determine the radius of the Earth.

The road along the Nile River (green) between Syene and Alexandria in Egypt had to be walked to measure the distance between these towns.

IS SPACE FLAT OR CURVED?

If we could draw grid lines across the universe in the horizontal and vertical dimensions, we could determine whether space is flat or whether it is curved in any number of ways. By 'flat', cosmologists do not mean flat like the top of a 2-dimensional table, but rather they mean in a Euclidean sense where a triangle sitting in any orientation would have its angles add up to 180°. If a universe has a flat geometry, space could go on indefinitely far beyond what we currently see in our observable universe. Alternatively, a universe in which space is curved outward like a saddle would also have no edge, and it too would go on forever. However, if the universe has a positive curvature similar to a sphere, a football, or a donut, then it would have a finite volume, but it would not have an edge - just like the surface of a ball or the surface of the Earth has no edge.

For the universe to have expanded from a single point, it would have to be flat. At present, our universe appears to be flat with no curvature of space. It has been measured by cosmologists to an accuracy of at least 1%. However, if the universe is say, billions of times larger than our observable universe, then 1% accuracy would not be enough to know if space was flat or curved. This level of accuracy would be like an intelligent ant trying to prove that the Earth was flat by measuring the sum of the angles in a triangle to be 180° when the sides of the ant's triangle were only a kilometer long. This length would

seem very long to the ant, so it would think that the angles were exactly 180°. But over that small distance compared to the size of the Earth, the accuracy required to know that the Earth was a sphere would be far beyond the ant's ability to measure the angles to the accuracy required. It's the same for us: from our current observations that extend across the observable universe, space appears to be flat, so it could be infinite. But if the dimensions of the Greater Universe are enormously greater than our observable universe,, then space may show no curvature at the accuracy we can measure, so it could be curved.

Our universe could have a geometry that takes the form of a torus, or any number of other forms. But this is impossible to visualize in an illustration because the geometry of the torus would occur in every direction. A toroidal universe would be curved to different degrees in different directions depending on which way one was traveling. Because space and time are intertwined, this might mean that the passage of time might also be different in different directions.

Illustrated here are three simple space-time geometries for the universe. The top one has a closed, positively curved geometry, the middle one is open and is negatively curved, and the bottom one is flat. Aside from these geometries, there are many other curved geometries that are more complex.

> *Top: If the curvature of space is positive, then our universe is closed and possibly a sphere. In this case, an equilateral triangle will have angles that add up to more than 180°, as is the case on Earth's surface. Alternatively, if it has a football or egg shape, or it is donut-like, then space would be curved to different degrees in different directions. With any of these options, the volume is finite.*

> *Middle: If space is negatively curved in two directions, then the sum of an equilateral triangle's angles will be less than 180°. This is the case with a saddle shape. With this geometry, space could go on forever, so it would be infinite.*

> *Bottom: If the universe is flat in all directions, then the sum of the angles in a triangle will be exactly 180°. A flat universe would go on forever, so it too would also be infinite.*

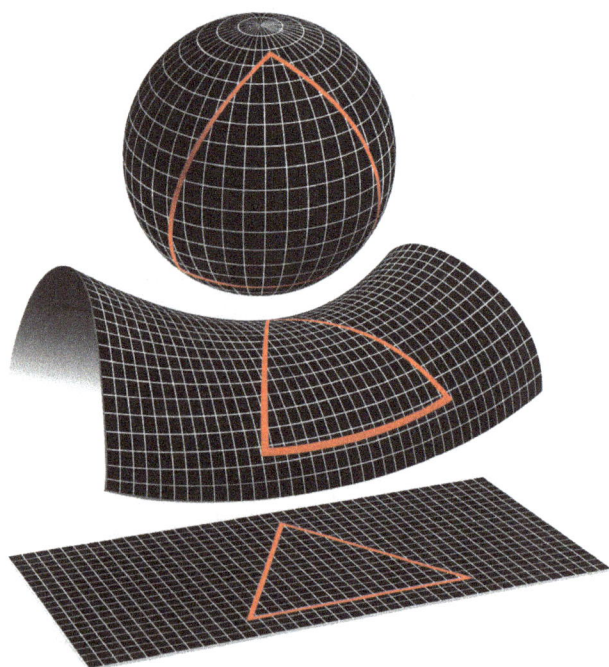

WHAT GEOMETRY DOES SPACE HAVE?

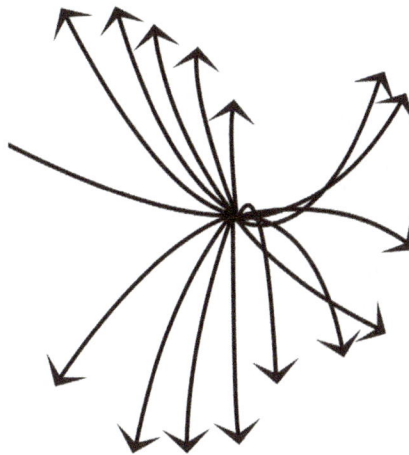

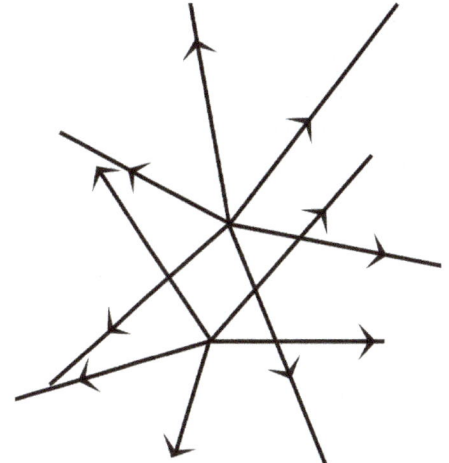

In each of the above universes, light beams will travel along specific paths as shown - assuming no distortion of space by bodies with mass.

Left: In a universe with a closed, positively curved, spherical geometry, light beams travel along circular paths no matter which direction they are pointed.

Center: In a universe with a negatively-curved saddle-shaped geometry, light beams would always curve outward traveling in negatively curved paths forever no matter which direction they are pointed. (Arrow on right cut off.)

Right: In a universe with a flat, open, infinite flat geometry, light beams will travel along straight paths forever no matter which direction they are pointed.

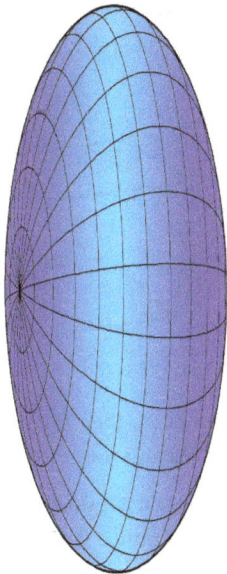

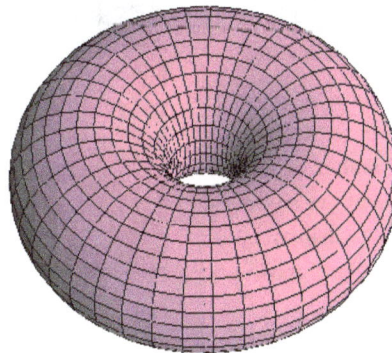

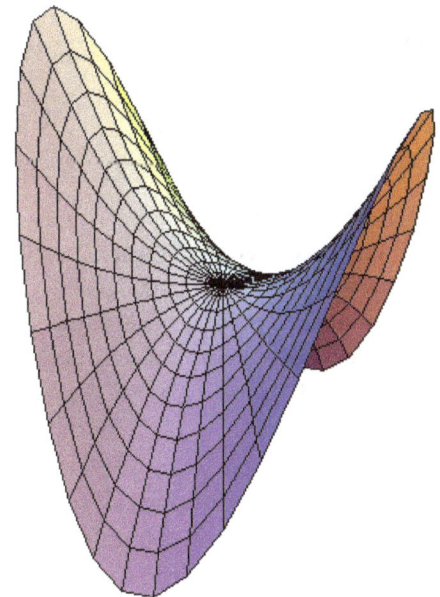

Left: **In a closed, ellipsoid (football or discus-shaped) geometry, light beams radiating from the poles are elliptical and of equal length. Those that are circumpolar like lines of latitude are circular but of different lengths. Lines that run at angles would be elliptical or spiraling and have different lengths. Space in this geometry is positively curved and therefore finite.**

Center: **In a closed, donut-shaped geometry, light beams that travel in horizontal or vertical circles would be circular but of different lengths. Those in the vertical would have the same length and those in the horizontal would have vary. Those that travel at angles to the grid lines would travel in ellipses or spirals of varying lengths. Space in this geometry is positively curved and therefore finite.**

Right: **In an open hyperbolic or paraboloid, saddle-shaped geometry, every light beam would travel along a curved open path to infinity. The curvature of space in this geometry is negative and infinite**

Let's imagine that we have a super-laser beam that can travel almost infinitely fast. In a positively curved space (without mass to distort it), the laser beam would come back from the opposite direction to that which it departed. In this case, its path would be circular, and space would be flat. All light beams in a spherical geometry would return after the same duration of time. Now, if some laser beams came back after different periods of time from different directions, this means that they have traveled along paths that have different lengths. This would mean that the universe's geometry would be either an ellipsoid, a donut shape, or something more complex. In a negatively curved geometry or a flat geometry, light beams will never return.

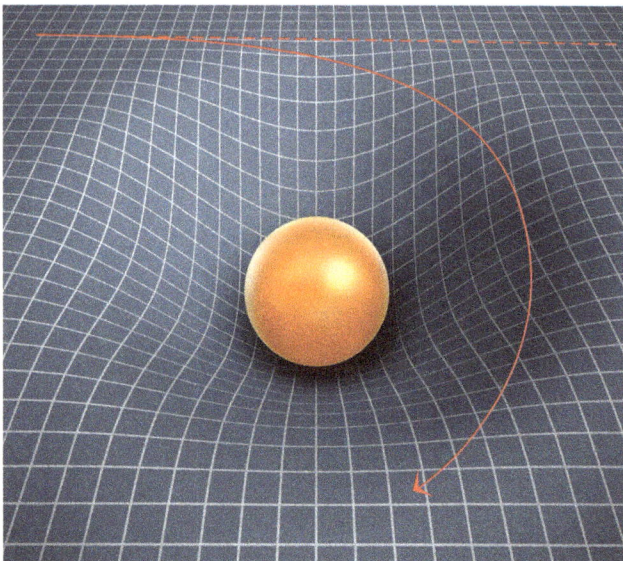

Light curves and twists as it passes through hollows in space that every mass creates. Light never travels in a straight line because it is always in the gravity wells of matter of one kind or another. Even in intergalactic space, there are slight depressions from galaxy clusters. (See Volume 2 Chapter 6, page 198 and Volume 3 Chapter 2, page 58 to see how gravity curves space-time.).

In reality, the laser beam test would not work like this because every mass in the universe would cause space to curve to some degree relative to the strength of its gravitational field. **Space is dimpled to various extents everywhere by the mass of galaxies on a large scale, stars on a medium scale, and planetary bodies on a small scale**. Because a light beam would curve in every 'gravity hollow', a light ray would never travel straight, so this test would never work. **Even though stars and galaxies look like their light is coming to us in a straight line, it isn't. All light beams twist and curve in every gravity field to some extent as they travel across**

space, so everything we see is distorted to some degree or other. Space is like the surface of an enormous golf ball, but the hollows are all different depths and widths and they are separated by wildly varying distances.

EXTRA DIMENSIONS

The mathematics of **String Theory** suggests that there are six extra spatial dimensions than the three we experience. String theorists think these extra dimensions might be curled up at infinitesimally small quantum sizes, and this is why we do not see evidence of them. (See page 14.) If there are extra dimensions in our macro universe which we are not aware of, then this would permit things to occur that would be far beyond our ability to comprehend them. But we do not see strange things happening, except in the quantum universe, so this probably means that there are no higher physical dimensions in our macro universe.

Because we can't experience more than 3 spatial dimensions, we could only see a 10-dimensional object's 3-dimensional shadow. It would have the most bizarre properties that would make it appear to change its form when looking at it from different positions. We would be able to see all around it and inside it from any position! It would be impossible to describe how it would affect light and space around it.

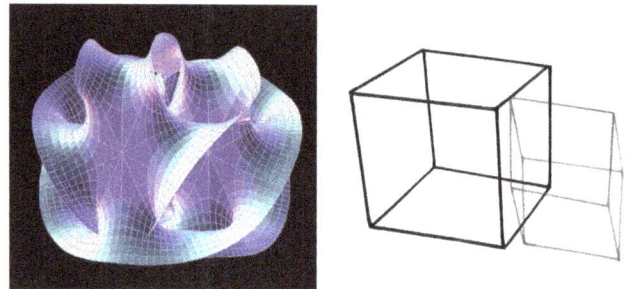

Left: **If the universe has 10 dimensions, then the shadow of a 10-dimensional object would look like this image.**

Right: **A 3D cube frame (black) throws a 2D shadow (gray). The shadow projected from a cube's frame onto a flat surface tells us very little about what it really looks like in 3D. We could never imagine what a 4D object would look like from its 3D shadow, let alone imagine what a 10D object would be like from its 2D shadow in the color illustration on the left.**

If a 10-dimensional object passed through our 3-dimensional space, we would only see a 2D cross-section of it at a time. To understand this, imagine looking at the shadow of a 3-dimensional wire cube projected onto a flat, 2-dimensional surface (as above). The shadow of the wireframe cube appears in 2-dimensions as a series of straight lines. When seen in 3 dimensions, the wire cube frame is far more interesting due to having depth that its 2-dimensional flat shadow does not have. If we rotate the 3-dimensional wireframe, we see its shadow lines moving in 2 dimensions, but this would not tell a 2-dimensional being what a cube is like in 3 dimensions. Similarly, a 10-dimensional object would be immensely more interesting, and intriguingly more complex than its 2-dimensional shadow shown in the diagram. If we could see a 10-dimensional object in 3 dimensions, it would appear like the most fantastic and magical thing. It would permit the impossible to happen. But we have never seen any indication of the effects of objects that have more than 3 dimensions. **This is strong evidence that higher dimensions do not exist, or it could be that we do not live in a region where any objects have higher dimensions.**

VISUALIZING HIGHER DIMENSIONS

To visualize higher dimensions, let's first imagine that there are 2D Beings that exist on a world called Flatland. This concept was brilliantly conceived by **Edwin Abbott** in his famous book '*Flatland*'. In Flatland, there are only two dimensions; one of length, and one of breadth. Flatlanders have no knowledge or perception of height in their 2D world. 2D Beings would consist only of lines or areas. They would not be able to see past lines, but they could infer an area by traveling around it. Imagine that there is a group of Flatlanders chatting and we pick one of them up and lift him into the third dimension. His friends would be aghast because they would see him disappear before their eyes! The Flatlander we picked up can now look down on his 2D world from above - something that he could never have imagined to be possible. We could make him fly around. Flight is a concept that a Flatlander would never have been able to comprehend. Going into the third dimension is a mind-blowing experience for this Flatlander. When we place him back down into his 2D world well outside his friends, they see him reappear as if by magic in another location! This is frightening to them. By going into the third dimension, this Flatlander experienced the thrill of seemingly unbounded possibilities, which would never leave his mind. If we were able to enter a fourth spatial dimension, it would have the same impact on us as the third dimension had on the Flatlander. We could put our arm out of a window and see it come back through a window on the opposite side of the room.

This concept is visualized very well in the movie '*Flatland: the Movie (2007) – A Journey of Many Dimensions*'. The voices of the characters are well-known actors. When my granddaughters were only 5 and 7 years old, they watched this thought-provoking film. They had a surprisingly mature conversation about it with me afterwards. At their young ages, I was surprised by how well they understood the concept and how enthralled they were with the idea. Despite their young ages, they were able to relate to how this could apply to us if we were able to go into a higher dimension. There is also a YouTube movie '*Flatland - The Wonderful World of Graphene*' that explains this concept, but it is not as good as the movie.

THE WEIRDNESS OF THE QUANTUM UNIVERSE

The closer we look at the quantum universe, the more it doesn't seem to make sense. In the quantum realm there are a number of inexplicable weird properties - the **wave-particle duality of matter**, the Heisenberg uncertainty principal, the strange properties of light, **quantum entanglement** where entangled particles have the same properties even when they are separated by great distances, the curious ability of **time and space being able to contract to nothing**, and the even weirder physics inside a black hole.

Quantum physicists discovered another strange property in the quantum universe. It's called **quantum tunneling**. This phenomenon occurs when an electron passes through a barrier that it should not be able to move through or go over. This occurs in nuclear fusion. It is utilized in quantum computing, the scanning tunneling microscope, and the tunneling diode. It occurs due to the wave-like properties of particles.

Thad Roberts has proposed in his book, '*Einstein's Intuition*' that strange quantum effects could be explained by there being more dimensions. A particle can be located by being at three spatial coordinates or dimensions, i.e. length (x), breadth (y), and height (z), plus one of time. But Roberts suggests that particles might also move in another dimension(s) that we are not aware of.

Quantized Units

Space and time must be quantized i.e. they must have a basic unit that cannot be divided – just as a pixel in a photograph cannot be divided. If we were to keep cutting any bit of gold in half, we would eventually get to having only one atom of gold. If we cut that in half, we would no longer have an atom of gold, so **matter is quantized**. Physicists are sure that **space and time are also be quantized**. If so, quantized distances in space and intervals of time are determined by the number of the quanta between them.

By having the x,y,z positions this permits us to identify a single quanta of space. An object cannot move around without changing its x,y,z positions. But space and time are distorted by matter. The more matter there is, the more space and time contract. When matter is illustrated as a hollow in space to indicate the curvature of space, the 'hollow' is not occurring just in the dimension of height, but in all dimensions. This could explain quantum tunneling and the Quantum Eraser experiment. Look this amazing experiment up on the Internet.

If we quantize the fabric of spacetime, then all units of measure can be reduced to 5 basic units – length, mass, time, ampere, and temperature. All the constants of nature are made up of these 5 numbers. Each unit has a value of 1. There are two other numbers that represent the limits of curvature. The diameter of the curvature of space is represented by Pi, while the depth of the curvature is defined by a number called Zhe. The reason there is a maximum number for these values is because space is quantized. So it can't go to infinity.

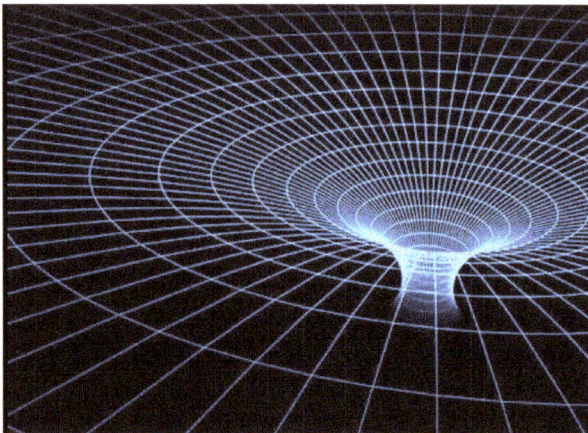

The curvature of space in a gravitational field appears to warp space in a dimension beyond the three spatial dimensions we are aware of. This extra dimension could explain strange quantum effects and the curvature of space.

There are 7 numbers that represent everything. This means that the geometry of space determines the values of the equations for these 7 constants. Dark matter and dark energy could be consequences of this geometry. (For more on other dimensions, watch the TED lecture '*Visualizing Eleven Dimensions*' by Thad Roberts.)

CHAPTER 10

MERGING GALAXIES

Galaxies have enormous gravity wells. When they come close to one another, they whip around each other in their combined gravity hollows. Their spiral arms become highly distorted as they are flung around. This often causes one arm to be drawn far away from the rest of the galaxy to form a long filament, before it is drawn back into the galaxies' gravity well.

During the merger of the cores of galaxies over a few bilion years,their supermassive black holes go into orbit around one another and eventually merge into one huge black hole. If the merger consists of two or more spiral galaxies, then the spiral arms of the new super-galaxy typically have renewed star birth due to clouds of gas and dust in each galaxy colliding with one another and becoming deneser.

The galaxies in some mergers can be rotating in different directions, so this can cause some interesting effects whereby the inner spiral arms are moving in one direction while the outer ones are going in the opposite direction. Spirals galaxies can merge with ellipticals. There can be up to several galaxies involved in a merger. Some galaxies pass right through other galaxies forming ring galaxies. The far outlying regions of some merging galaxies can end up forming a huge halo around the newly formed galaxy. Merged galaxies share their globular star clusters, but some are flung off into intergalactic space. (Refer to Volume 3 Chapter 8, page 388 for HST images of galaxies merging). Video computer simulations of galaxy mergers are available on YouTube. They are fascinating to watch.

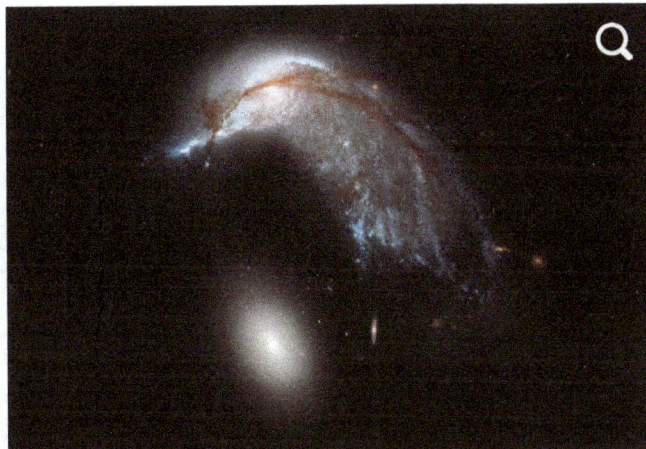

Left: **The Mice Galaxies NGC 7714 and NGC 7715 are in the process of merging.** Credit: HST NASA

Right: **The merging Penguin and its Chick Galaxies are also known as Arp 142. The penguin is the very distorted spiral galaxy that is at the top. It is being torn apart as it orbits the elliptical galaxy below it (the egg).** Credit: HST

Galaxies like NGC 2623 were described decades ago as 'peculiar' galaxies because astronomers back then did not have telescopes with enough resolution to see that these objects were galaxies merging. Grazing blow mergers can have several destructive passes before they merge into one large galaxy. Credit HST, NASA

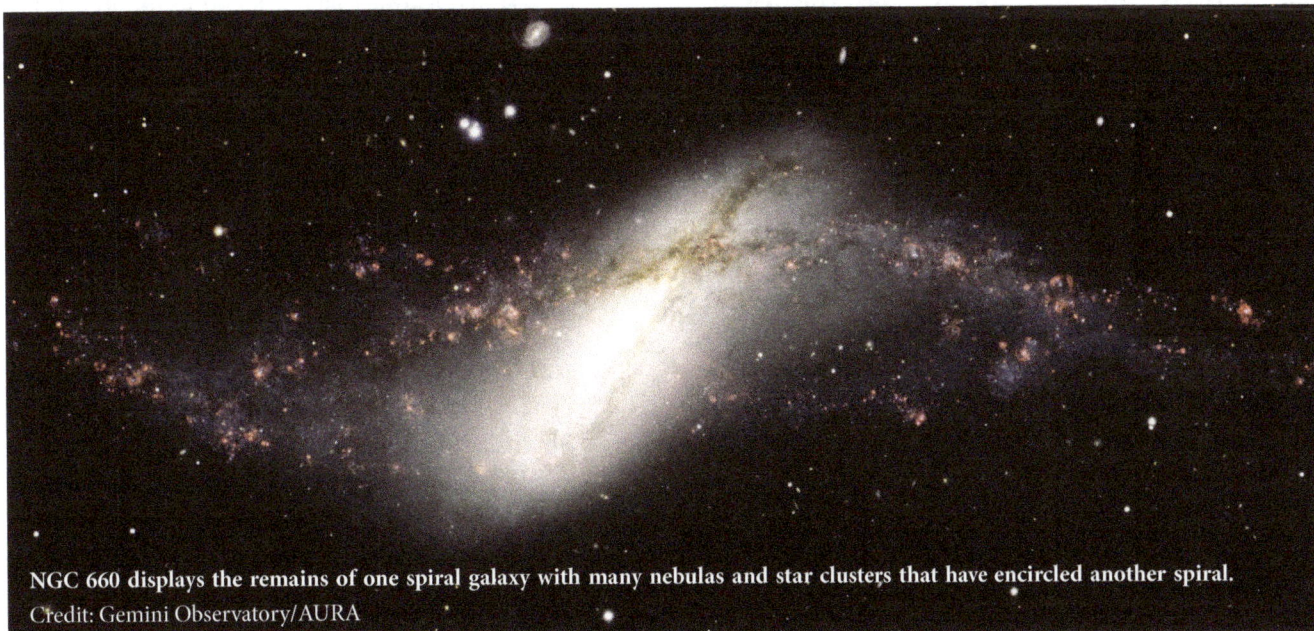

NGC 660 displays the remains of one spiral galaxy with many nebulas and star clusters that have encircled another spiral. Credit: Gemini Observatory/AURA

These two dramatic merging galaxies are known as Arp 273, or UGC 1810. The outer arms are very rich with star birth. The inner galaxy has distinct dust lanes. Credit HST, NASA

HOW GALAXIES MERGE

Given how much matter is thrown off galaxies when they interact with one another, there must be a lot of it strewn between them, around them, and far from them. Two teams of astronomers have now confirmed that there is around twice as much matter in intergalactic space as previously thought. This reduces the amount of dark matter required. It also opens up the possibility that dark matter may be ordinary matter that we do not yet know how to detect.

THE STAGES OF MERGING GALAXIES

1. Over billions of years, scores to hundreds of galaxies will merge to form a cluster of galaxies. **Giant ellipticals, which are formed from many smaller galaxies, form at the centers of these galaxy clusters.** Giant ellipticals have masses that range from around several to more than 10 average-sized galaxies.

2. Galaxy clusters that are within one another's gravitational influence will merge to form a supercluster. These can contain thousands of individual galaxies of all sizes including a number of very massive ellipticals. These central elliptical galaxies eventually merge to form giant, super-massive elliptical galaxies. Around these, there are hundreds of average-sized galaxies with tens of thousands of dwarf galaxies and a hundred times this many globular star clusters. When galaxies merge, some of their globular clusters are flung out of the cluster into intergalactic space. During mergers, the central, supermassive black holes of each galaxy merge to become a much larger black holes. These produce fantastically large jets. (See Volume 3 Chapter 2, page 94 onwards)

3. Superclusters that are within one another's gravitational grasp, will eventually merge to form a **super-massive supercluster**. Some of the central elliptical galaxies in each supercluster will merge to create **monster mega-elliptical or mega-spherical galaxies.** These can have masses of up to a thousand times that of an average-sized galaxy like our Milky Way.

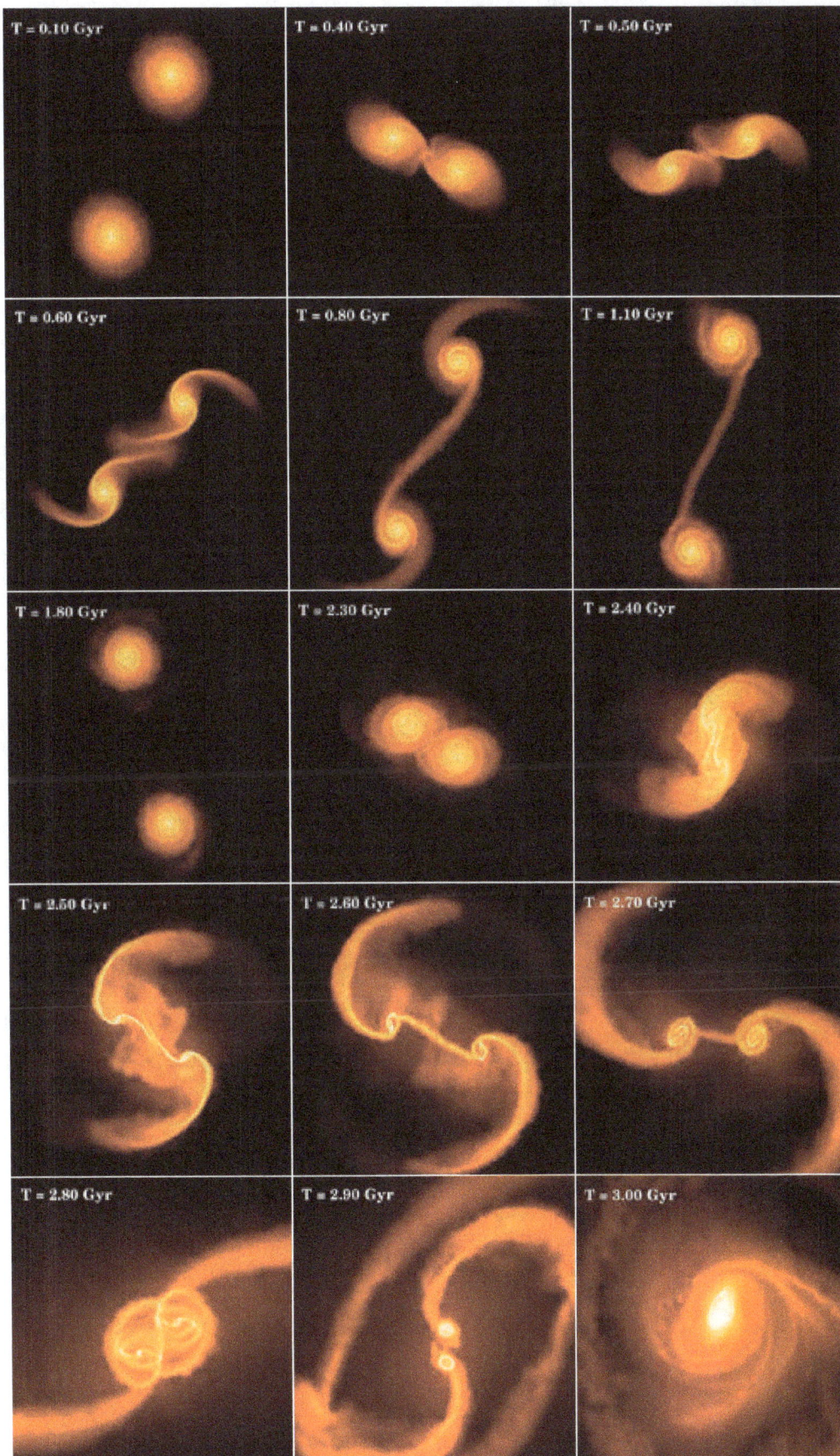

Here we see video frames from a computer simulation that shows the process of two galaxies merging into one huge galaxy. In each box, the time frames are given in billions of years (Gyr).

When many galaxies merge, they form giant, supermassive elliptical galaxies, like the one seen here at the center of this very distant supercluster. It has a multiple core due to two or more giant elliptical galaxies merging. There is a faint jet projecting from its top left. The jet is emanating from the galaxy's central supermassive black hole. Note the large number of other smaller massive elliptical galaxies and smaller spirals swarming around it. In time, all these galaxies will eventually merge with the central supermassive elliptical. The arcs are distorted images of very distant galaxies that lie far beyond the giant elliptical. Its gravity is acting like a telescope lens to focus, distort, and magnify very distant galaxies that are directly behind it. This effect is known as gravitational lensing. It is a good thing that our galaxy is not near such a bright elliptical galaxy because our night sky would be as bright as day. Credit: NASA HST

Here we see the cluster Abell 1413 lying 2 billion light years away. It has a huge super-elliptical galaxy with multiple nuclei involved that is around 7 million light years long. Many other large ellipticals and spirals are in orbit around it. There are some 300 galaxies in all in this cluster, but most are too small to be seen as anything but specks at this scale. Credit: NASA HST

Over hundreds of billions of years, many supermassive clusters will merge to form a single, gigantic, mega-massive elliptical 'galaxy' made of tens of thousands of galaxies and hundreds of thousands of globular clusters that all become a part of the mega-galaxy. As the universe ages, almost all galaxies will eventually become a part of one of these monsters. Illustration by NASA

WHAT IS THE GREAT ATTRACTOR?

Enormously massive collections of clusters of galaxies create deep gravity wells in space. They are so deep that they draw in galaxies from hundreds of millions of light years away.

When the recessional velocities of galaxies are measured, (see Volume 3 Chapter 8) it becomes clear that in our region of the universe, there is an overall flow of galaxies in the direction of the **Norma-Hydra-Centaurus supercluster**. This supercluster extends across a region between 150 and 200 million light years away. It has a mass equivalent to tens of thousands of average-sized galaxies.

Around 400 million light years beyond this supercluster lies the even more massive **Shapley supercluster**. It consists of a group of very massive galaxy clusters that have Abell designations, named after George Abell, the astronomer who discovered them. The Shapley cluster accounts for more than half of the gravitational pull of these two superclusters, so the Norma-Hydra-Centaurus supercluster is being drawn towards it. The combined mass of these two superclusters has been coined **The Great Attractor**

because its gravity is drawing in other clusters and small groups of galaxies from up to half a billion light years away in every direction! This includes our Milky Way and our Local Group of Galaxies. (See Volume 3 Chapter 8, page 293.) It's astonishing that gravity, a weak force on smaller scales, can affect enormous bodies of matter on these very large scales.

Most superclusters are too distant for our present-day telescopes to detect them. There is no doubt that many more superclusters exist beyond those that we can detect. Galaxies that are far away, close to the edge of the observable universe, are not seen as being in clusters like they would be today, because we are seeing them as they were when they first formed before they had time to clump together.

There will be other 'Great Attractors' across our observable universe, and incalculable numbers of them in the Greater Universe beyond. Some will be even larger and more massive than those that we can see at present because they will have had billions of years more time to condense. In the future, as superclusters merge, there are likely to be superclusters of superclusters that possess colossal gravity wells.

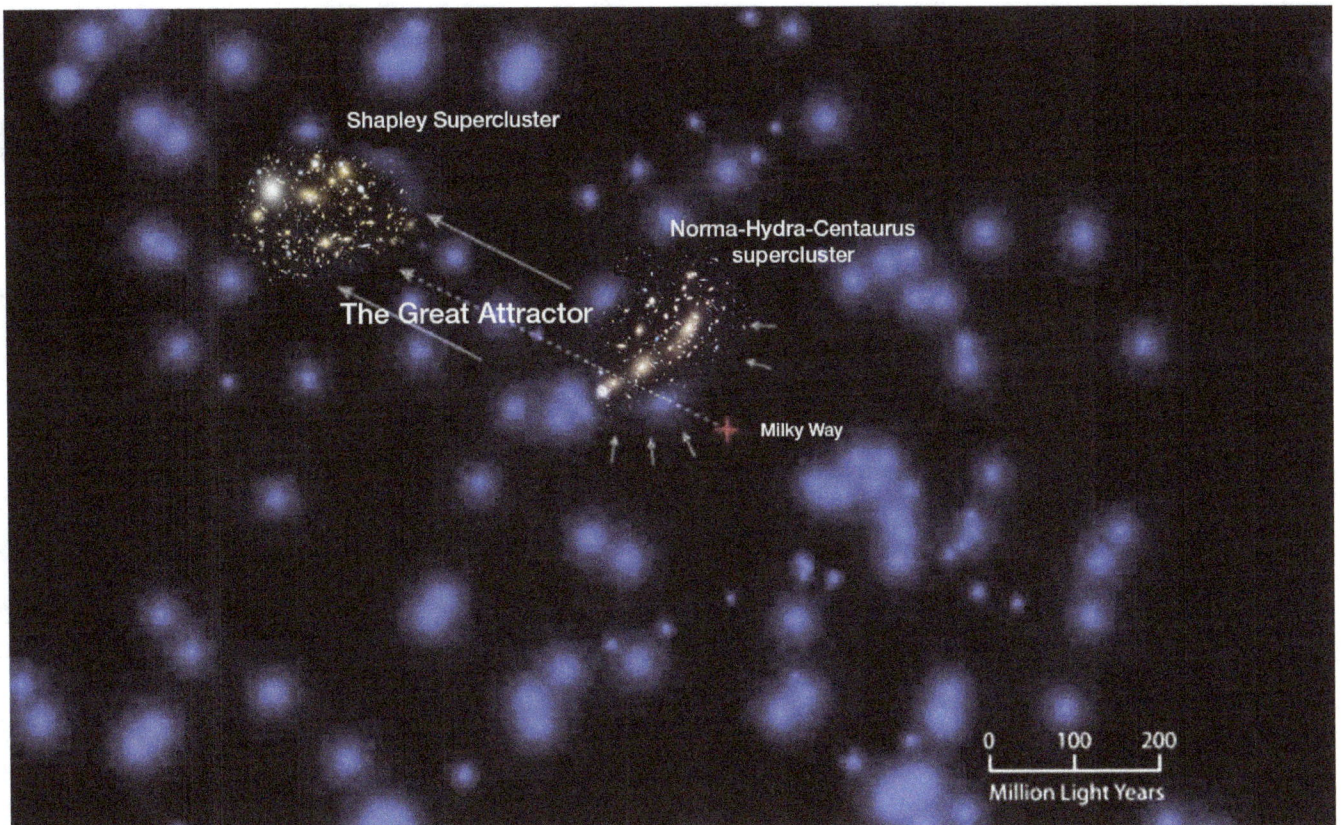

All the blue spots in this diagram are clusters of galaxies within 700 million light years of the Milky Way (red cross). The closest super-massive cluster of galaxies to us is the Norma-Hydra-Centaurus supercluster lying between 150 and 200 million light years away. Beyond that is the even larger and more massive group of galaxy clusters known as the Shapley supercluster centered on 650 million light years away. These two super-massive clusters are attracting numerous other galaxies and clusters toward them. This includes our Milky Way. These two superclusters have been named the Great Attractor.

This amazingly densely populated supercluster is a member of the Great Attractor Group of superclusters. It displays many supermassive elliptical galaxies formed from numerous mergers. Note the number of distorted images of distant galaxies that have been gravitationally lensed to appear as arcs. In this extraordinary image of a supercluster of galaxies, many Einstein arcs of distant galaxies are visible. They are caused by the gravitational fields of the major galaxies in the cluster acting like a lens to magnify very distant galaxies and distort their shapes. A galaxy may be lensed more than once Credit: HST, NASA

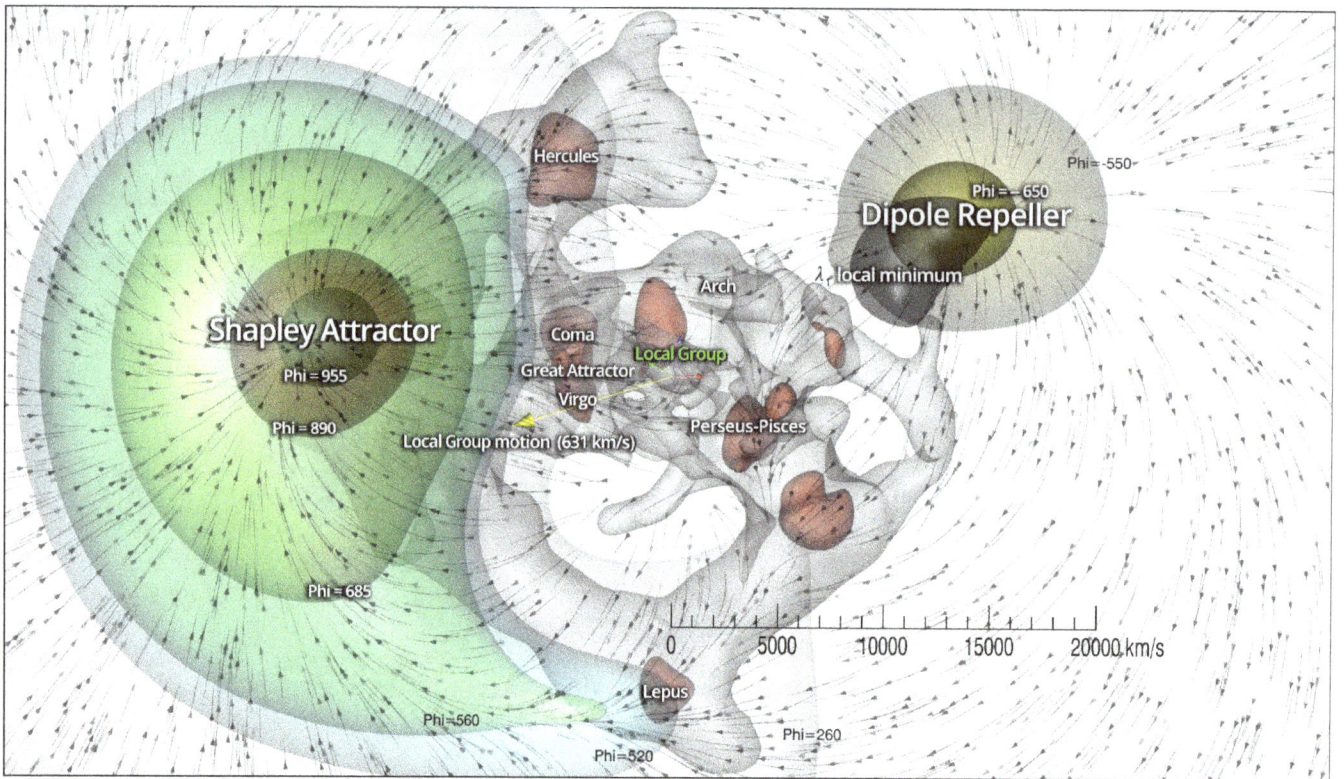

Our Local Group, the Virgo Cluster and Coma Cluster are all being drawn towards the supermassive Shapley cluster of superclusters. Originally, the Coma-Virgo supercluster was known as the Great Attractor until it was realized that the much larger Shapley mega-clusters (green on the left) was what was attracting all the smaller galaxy clusters to the right of it. Others smaller superclusters of galaxies are shown in red. Areas of the same color are regions with the same gravitational values. Astronomers have discovered an anti-gravity force in empty space where there are no galaxies. This force is known as the 'Dipole Repeller' (yellow). It could be filled with dark energy. The Repeller force is pushing galaxies towards the Shapley supercluster at a velocity of 2 million km/hr. Credit: Hebrew University of Jerusalem.

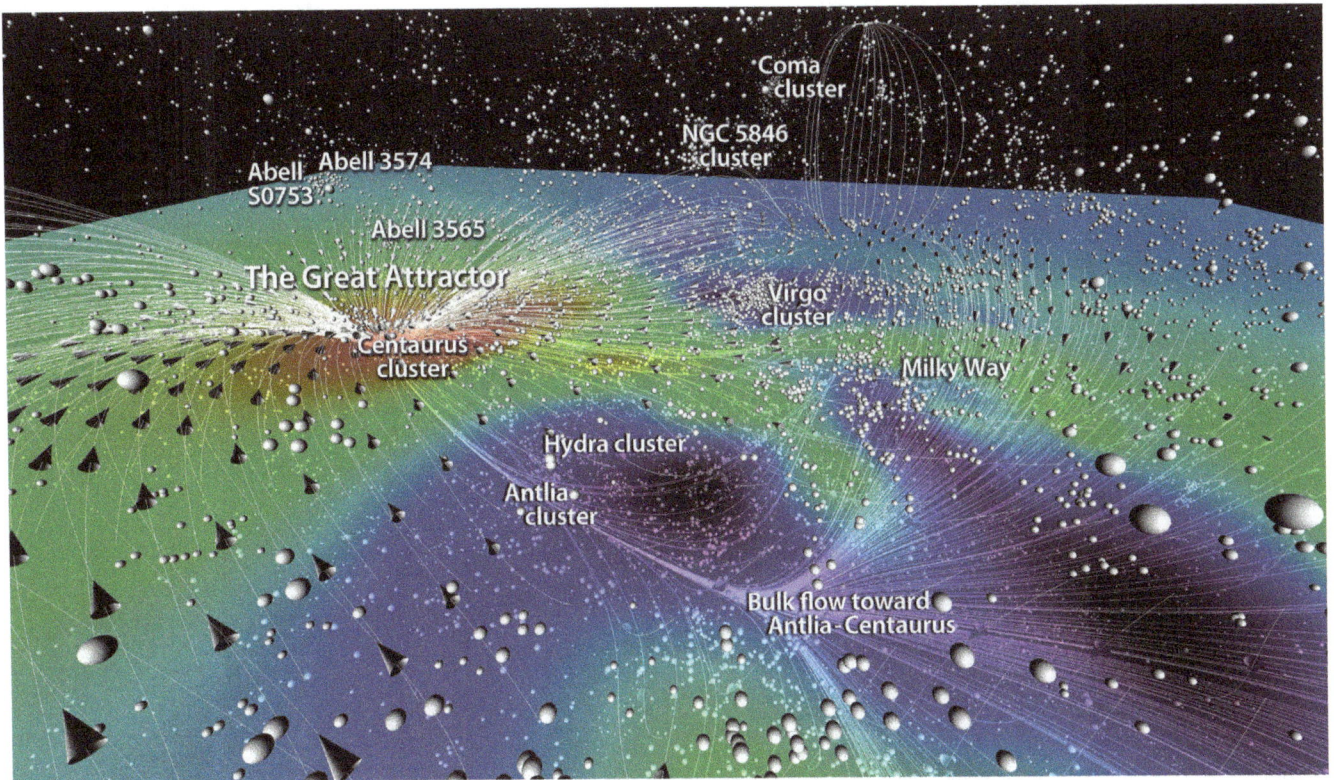

In this close-up view of our local universe, individual galaxies and clusters of galaxies are seen being drawn towards the Centaurus cluster with the Shapley supercluster behind it.

In this diagram, galaxies (gray spheres) are flowing along gravitational 'currents' (black arrows on white lines). The red region is the Great Attractor's strong gravity well that is drawing galaxies into it. The dark purple and dark blue regions are large voids devoid of galaxies. These are presumed to be regions of dark energy that act as anti-gravity to repel galaxies. What this diagram represents is similar to a landscape where rain runs off mountains (the dark purple areas) into lakes (the orange-red areas). Green regions are where moderate gravity fields exist and space is rather flat.

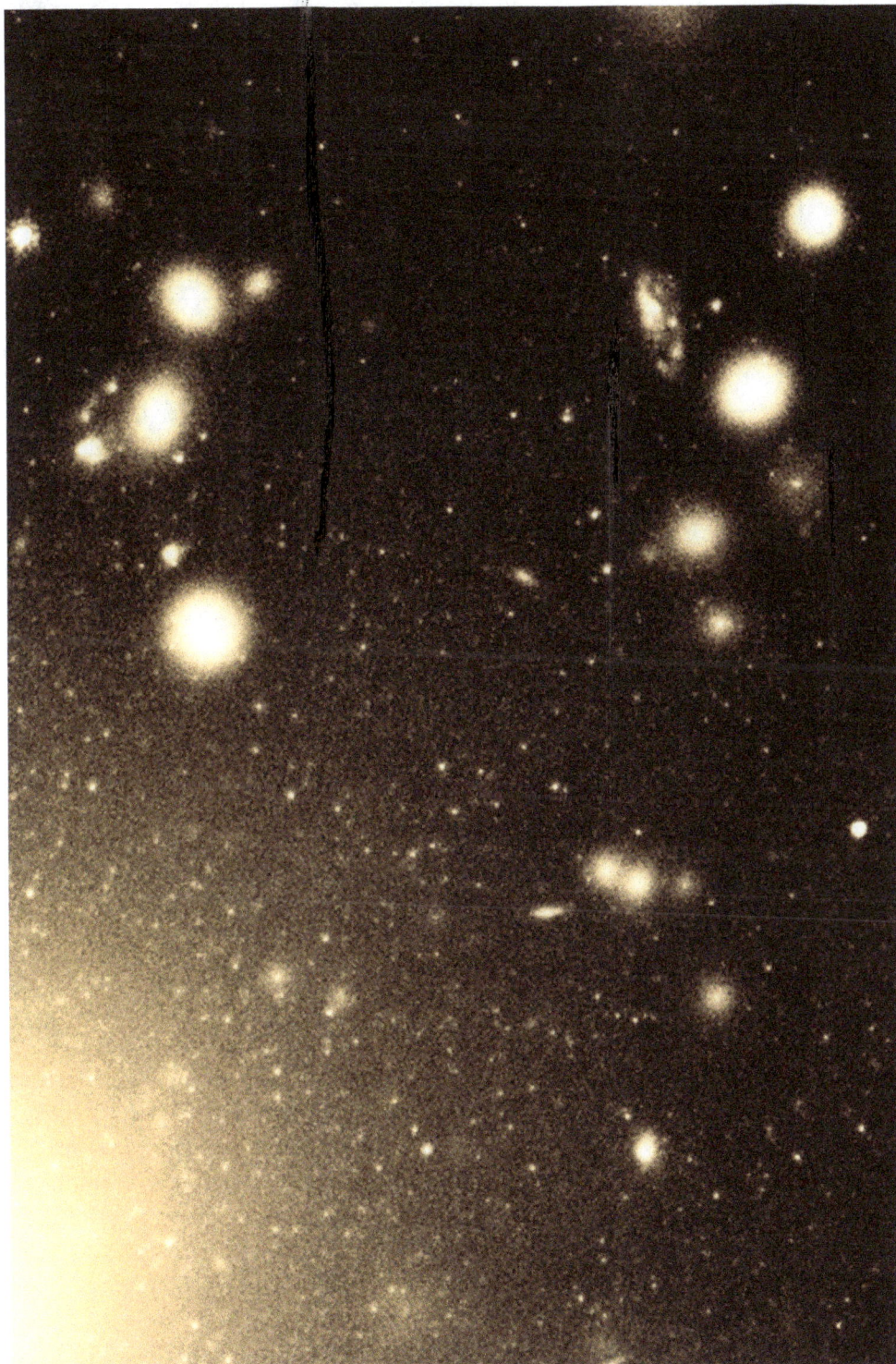

Here we see the Abell 3558 supercluster of galaxies. This wide-field view is but one part of the Shapley supercluster of galaxy clusters. The image contains many very large elliptical galaxies that formed after numerous average-sized galaxies merged. There are thousands more smaller galaxies involved that are too faint to show at this scale, and at such a huge distance. We are seeing these galaxies as they were 650 million years ago. Since then, they would have become far more compressed. Note the fine faint blue gravity lensed galaxies in the lower half. Only four blue stars are seen in this picture (top right corner with spikes). Every other speck is a galaxy.

This is a monochrome enlargement of just the inset box in the left image. In the lower left, it shows the outer edge of just one of the super-massive, elliptical galaxies. The field is filled with countless large dwarf galaxies as well as super-large globular clusters seen as grain. There are also many average-sized spiral and elliptical galaxies, together with seven large elliptical and spherical galaxies. Each minute faint speck is a large, bright dwarf galaxy, or a super- sized globular star cluster. Thousands of average-sized and smaller globulars are too faint to be recorded at that distance. When you look into these images in high-resolution photographs, the number of galaxies is truly astounding. Credit: HST, NASA

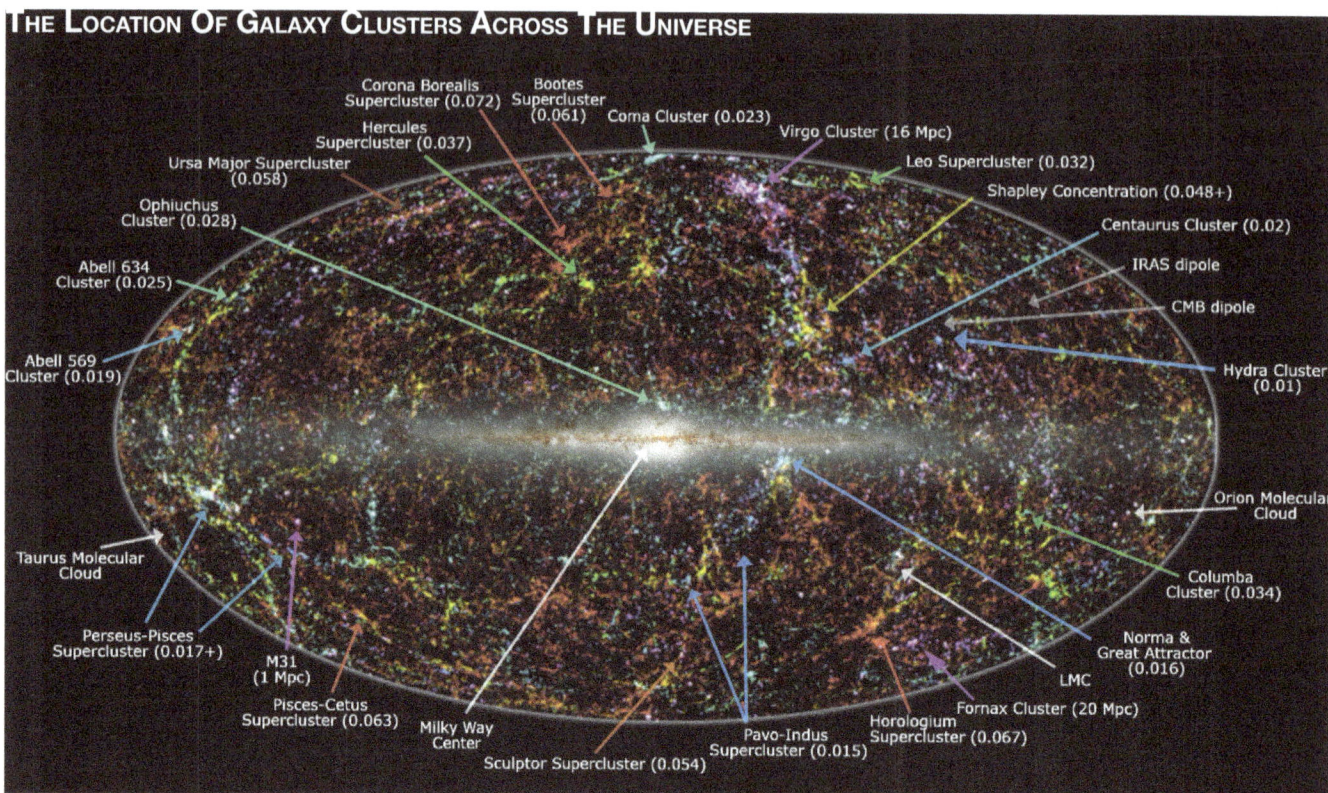

THE LOCATION OF GALAXY CLUSTERS ACROSS THE UNIVERSE

CHAPTER 11

THE FUTURE OF THE UNIVERSE

Neither space nor the objects in space are moving with the universe's expansion. It is the size and geometry of space-time that is changing its scale to make the universe larger. This can be likened to having paper spots glued onto a large balloon that keeps expanding as it is inflated. The spots do not change their size or move; it is the distance between them that keeps enlarging.

Astronomers think that **the universe should have no net mass**. This is because the energy that drives its expansion is expected to equal the gravitational attraction of the mass of all its galaxies. If this were the case, then gravity would eventually overcome the expansion causing it to stop. The universe would then begin to contract ever faster into a 'big crunch'. A new singularity would form and possibly a new Big Bang might occur. This concept is called the **cyclic universe**. If the universe continues to expand, space will become more and more negatively curved causing it to expand ever faster. If this occurs the universe would eventually tear itself apart right down to the level of matter.

We do not experience an anti-gravity force locally in our Local Group of Galaxies but there does seem to be because an anti- gravity force at the centers of the voids between galaxy clusters..

Presently, galaxies and galaxy clusters have enough gravity to stop space expanding locally, however, in the distant future, this may not be the case.

WE ARE VERY PRIVILEGED

If there are intelligences like us that evolved hundreds of billions of years in the future, they will not see numerous individual galaxies, and galaxy clusters strewn across the universe and receding from one another as we do. All galaxies beyond their own cluster will have moved beyond their observable universe. The galaxies in their cluster will merge with one another to form a single mega- massive mega-galaxy full of old, dim orange dying stars. No new stars would have been born for a long time because all the hydrogen will have been consumed long ago by its existing stars. Intelligences in this distant future will see no evidence of the Cosmic Microwave Background Radiation because it will be far beyond the horizon of their observable universe. Intelligences in these galaxies will have no way of knowing that there is a much Greater Universe. They would not know where the universe came from and they would not know that the Big Bang occurred. They would therefore not know how the universe evolved and they would not know how old it is. They would see no evidence of stellar evolution or the birth of stars. And nor would they know that space beyond their mega-galaxy is expanding. To them, their galaxy would be the entire universe.

On the other hand, intelligences that evolved in the early universe would only be aware of a much younger universe where galaxy clusters are much younger, smaller and much closer to one another than they are today. They would see far more galaxies than we can see because all the galaxies that have moved beyond our horizon due to the universe expanding, would be inside their observable universe. They would see only small groups of galaxies starting to form because large clusters would not have had time to come together. Intelligences in the early universe would not know that galaxies will become so separated from one another that they will only see their own galaxy. It would be too early in the universe's evolution for them to know how long most stars will live or how stars and planets will evolve in their later life.

It is quite remarkable that we are lucky enough to exist at a time where we can not only see back to the birth of the universe, but we can also look forward to see how the universe is most likely going to evolve in the distant future. We appear to be living in a very special time of the universe's lifespan. Is that luck, or by design? Or, are we deluding ourselves because there is so much that we do not know?

THE DEATH OF THE UNIVERSE

It appears that the universe will die, as all things in it do. So far, there seems to be two possibilities for its demise:

1. **The Big Rip** - the increasing expansion of space will eventually tear the space-time continuum apart causing galaxies, stars, and ultimately all matter to disintegrate.

2. **The Heat Death** - this is the death of the universe through stars losing their heat. All the available hydrogen in the universe at present, will be consumed by stars that will stop giving off heat

and light. Their heat will be spread evenly over the whole universe causing it to become so cold that it will only be a few degrees above absolute zero. The universe will become a totally dark carcass with only black, dense cinders of what were once stars.

Like everything on Earth, elements in the present universe are recycled to create new stars and new life, but this will not occur far into the future when there is no hydrogen left.

Of course, we may make discoveries in the future that could change both of these outcomes.

CHAPTER 12

THE UNIVERSE'S TWO OPPOSING PROPERTIES - ENTROPY & ORDER

Just as there are many opposing forces in nature, such as the positive and negative forces of electricity, there are two important forces, without which the universe would not evolve. One is **disorder**, known as entropy, and the other is **order** which drives increasing complexity. Disorder results from the natural process for everything in the universe, including the universe as a whole, to ultimately move towards disorganization and decay. For instance, entropy causes all stars to eventually burn out and die - just as it causes all life forms to die and decay. Order grows in pockets to form such things as stars and life forms. These pockets of order oppose disorder for the time they exist. They create new structures and increasing complexity. But in time, everything ages and entropy wins out in the end. For this reason, like all life and non-life, it appears that the universe will eventually die too like everything in it. Perhaps, very high levels of highly intelligent order may stop this, or, it may be that highly intelligent consciousness evolves to a state that is above requiring a physical universe.

The **Second Law of Thermodynamics** simply states that in a closed system (i.e. a container, a life form, or the universe), heat can only flow in one direction - from a hot region to a colder one - and never the other way (unless energy is expended to make this happen, as in a refrigerator). Abiding by this law, disorder will cause the heat concentrated in stars and planets to eventually spread out across the vastness of the universe over trillions of years. There will be no remaining regions with differences in heat energy, so heat will not be able to be transferred from a hot region to cold one. As a result, matter will not be able to interact with other matter, so no new stars will form, and no life will be able to exist.

If there was not the force of order to counteract disorder, then the universe could never have evolved. For the universe to evolve, pockets of high order must occur amongst the disorder. As order evolves, it becomes more complex. This occurs in molecular structures, astronomical bodies, cells, intelligence, consciousness, technology, and so on. Because of this, I contend that there is an underlying universal law, which stated simply says, '*the universe shall become more complex*'.

In a closed system, order leads to increasing complexity. When the complexity of molecular structures reaches the level of cells, we call this closed system 'biological life'. Life is a state that creates order within itself at the expense of *creating disorder (entropy) outside itself*. It does this by expending energy and excreting waste. **Life is a form of organized energy that incorporates complex systems of molecular order within it.** When life dies, its elements are released back into the environment to be recycled into new life forms. It appears that nearly all star systems throughout the universe will form planets that have complex molecular structures. On some worlds where conditions are suitable, very complex structures like DNA, cells, and, advanced lifeforms will evolve. But each star, planet, life form, and each cell will have a limited lifespan.

It now appears that life will soon be proven to be a natural evolutionary process that creates pockets of order throughout the universe.

In the Greater Universe outside our observable universe, there may be pockets of different types of disorder and order which we cannot imagine. In our observable universe, it appears that we may have pockets of life that evolve, but in parts of the Greater Universe pockets of life may be very rare. An analogy for this could be that rainforests have an abundance of different life forms, whereas dry, hot deserts and cold Antarctic wastelands do not. We have no way of ever knowing what conditions lay beyond our observable universe because information cannot come from beyond the horizon of any observable universe.

WHAT DRIVES THE UNIVERSE TOWARDS INCREASING COMPLEXITY?

In this section, I will suggest one factor that could be causing the universe to evolve to have continually increasing complexity.

Edward Lorenz discovered that chaos was actually an unsuspected higher level of order.

THE DISCOVERY OF CHAOS THEORY

The brilliant mathematician and meteorologist **Edward Lorenz** is famous for developing **Chaos Theory**. Before Lorenz discovered that the laws of chaos caused many systems to appear chaotic, chaos represented a mathematical abyss to both physicists and mathematicians because it appeared to have no order to it. But **Lorenz discovered that chaos was actually a *higher level of order that produced greater complexity***.

Lorenz found that **the smallest changes in a system could multiply and lead to major changes**. This became known as the '**butterfly effect**'. **This is the universe's way of driving continual change in every aspect of its existence, causing the universe to continually evolve towards ever greater complexity.**

In the late 1950s, Lorenz was working on trying to improve weather forecasting. In 1961, he used a primitive computer to predict the outcome of the weather. His program for predicting the weather was a 12-component model that measured temperature, wind speed, air pressure, humidity and so on for different altitudes in the atmosphere over a region covering Britain,

France, Scandinavia, and the North Sea. He calculated predictions for what the weather would do over the coming days using initial conditions to six digits. On one occasion, he ran the program a second time, but this time, to reduce computing time, he decided to round off the initial conditions to three digits instead of six. He assumed that it would make no difference to ignore such minuscule differences in the initial conditions, but to his amazement, over five days this produced a completely different outcome! The graphs matched for the first few days, but as the days went on, they drifted apart. Instead of it being a fine day over Paris in 10 days' time on his first prediction using 6 places, the second prediction using only 3 places forecasted cyclonic conditions! Most people would have shrugged this off as some quirk in the program, but Lorenz investigated what the reason for this could be. He discovered that very small differences in the first 6 decimal run kept multiplying to cause major changes that did not appear in the 3 decimal run. This became a major discovery that led Lorenz to develop **Chaos Theory**. His theory became the key to understanding why numerous chaotic systems in the universe work the way they do.

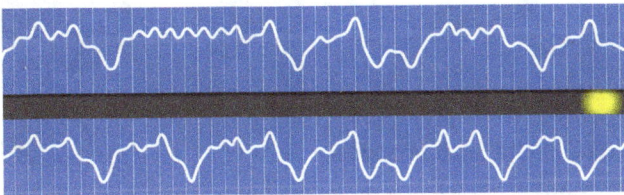

Lorenz's top graph plotted the weather over a month using initial conditions rounded off to six places, whereas the bottom graph used initial conditions rounded off to three places. Note how the outcome stays in sync for about 5 days. After this, it changes dramatically due to very small butterfly effects multiplying. Halfway through from day 14 to day 21, the graphs become fairly synchronized again, but then they drift apart once more to produce completely different outcomes. If 12 or 20 places of accuracy were used in calculating the predictions, this would produce a different outcome again.

Chaos Theory showed that uncertainty in chaotic systems increases exponentially with time. Chaotic systems do however have limitations that cannot be exceeded. For example, while the temperature changes all the time across the globe, the average global temperature will never reach 60°C or drop to cause a major ice age under the present astronomical conditions for the Sun and the Earth. Current conditions should exist for at least another thousand years barring an asteroid impact or a major eruption of a supervolcano such as Yellowstone.

Lorenz had discovered that all chaotic systems are affected by the Butterfly Effect. It got its name from the possibility that under the right conditions, the mere flapping of a butterfly's wings in one region of the Earth could potentially cause a hurricane over another region in the weeks ahead due to Chaos Theory.

Lorenz had discovered that local weather cannot be predicted with any accuracy for any more than 5 days because too many minute changes are always occurring, and these keep multiplying to change the final outcome. Beyond 5 days, the chances of a prediction being right diminished rapidly, so much so, that by 10 days it was down to pure chance for most regions on the planet.

Despite computers becoming thousands of times bigger and faster than Lorenz' primitive computer and there being many weather satellites monitoring the globe, the Butterfly Effect is the reason that weather forecasters still can only make reasonably accurate predictions for only a few days ahead. Some errors still occur even in that small window.

All 50 computer models used for predicting global warming have always turned out to produce false outcomes due to the effects of Chaos and its Butterfly Effect. Because it is impossible to predict the weather for any location with any certainty for more than five days ahead, predicting what the climate might be a decade from now, let alone a century, is simply impossible.

The biggest advance for predicting global climate will come from programs that integrate the astronomical effects that the Sun's weather has on the Earth. This would include the intensity of the Sun's storms, if a coronal mass ejections is headed towards our planet, the intensity of the solar wind, and the intensity of cosmic ray showers from deep space that affect the degree of precipitation that occurs in Earth's atmosphere. (See Volume 2 Chapter 3, page 85.)

Computer programmer **Michael Chermside** wisely states that with regard to predicting climate changes *"too much trust in computer programs can lead us astray. Invalid input, invalid assumptions, invalid equations, and mathematical chaos, all conspire to cause many pitfalls."*

He also says that *"an experiment that fails may, in the end, be even more useful than one that succeeds (because it) allows us to discover things about nature that we do not understand at present."*

A SIMPLE FORMULA DRIVES CHAOS

The butterfly effect can be described simply in algebraic terms by this basic formula:

$$\overset{\curvearrowleft}{X^2 + C} \Rightarrow X$$

In this equation, 'X' represents anything, and 'C' is a change of any kind on any scale. When X is squared and then changed by C, the outcome value of X is affected. This new value for X is fed back into the equation and squared again and a new change is added to produce a new value of X. This is repeated ad infinitum. This process produces never-ending change. The rate of change may be extremely fast or very slow depending on the value of C, and how fast the feedback system works. If C represents a large change, this will cause a rapidly changing effect. However, over time, very small changes can also multiply and cause large changes to occur.

AN EXPERIMENT THAT VISUALLY DEMONSTRATES THE BUTTERFLY EFFECT

C To see how dramatically the Butterfly Effect works, you may like to perform the following captivating experiment. In a dark room, connect a video camera to your TV or computer monitor screen. The camera should see just the blank screen. If the room lights are on, you will see an image of the screen repeating to infinity.

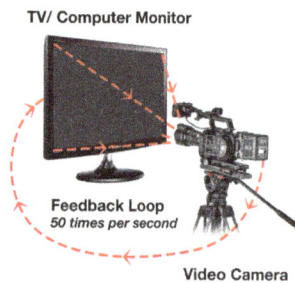
TV/ Computer Monitor
Feedback Loop
50 times per second
Video Camera

Turn the room lights off and hold a lit match briefly in front of the screen. The video vision will show the flame. Now, draw it away quickly so it is out of the camera's view and extinguish it. You'll notice that the image of the flame will remain despite the match no longer being there! This is because the initial image is continually being fed back into the circuit. But the image will look nothing like the flame! It will have quickly changed in a remarkable

way. Due to the feedback loop in this system, you will see a constantly evolving, complex pattern radiating out from the center of the screen. It will be reminiscent of computer-art video vision that simulates traveling through a 'space warp' in a science fiction movie!

The reason the image keeps changing is due to minute changes in the temperature of the electronics in the circuitry of the camera and the screen as well as other small chaotic things occurring in the electronics. There are distortions caused by tiny imperfections in the camera optics and from particles in the air moving between the camera and the screen. Each small thing slightly changes the image 50 times a second. The process is repeated over and over causing a cascade of changes from one frame to the next. This makes it look like a movie. Changing the camera angle very slightly will create a new set of initial conditions that will cause a different image to evolve.

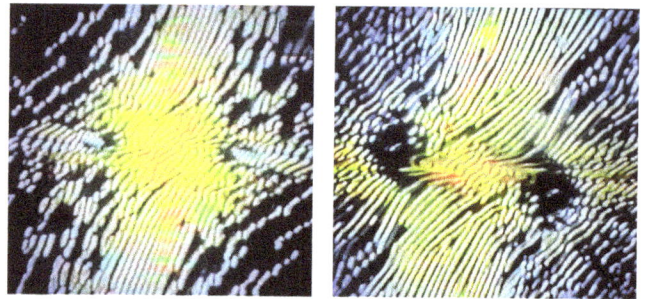

The butterfly effect causes the feedback video image to continually evolve every fraction of a second making the image continually change making it appear as if you are flying into it.

Once the effect is running, you can save it as a video and with a video projector, you can project it onto a wall at parties to create a great special effect! While the screen image changes endlessly, you will note that there is a limit to its degree of change. This experiment is a visual demonstration of how chaotic dynamical systems continually evolve and how their changes are limited to the boundaries that the system has. Butterfly effects like this are occurring everywhere across our world and throughout the universe at every level - from molecules to superclusters of galaxies.

Smoke rising from a cigarette in a still room follows classic chaotic behavior. The smoke rises in an orderly vertical fashion at first, but soon, minute butterfly effects such as the variable current or temperature in the air and in the rate of burning of the cigarette, cause the flow of rising smoke to become increasingly more turbulent and chaotic, the higher it rises. The same occurs with a flow of water. Seemingly insignificant changes in the initial conditions of chaotic systems multiply to produce unpredictable outcomes.

BUTTERFLY EFFECTS AT WORK IN NATURE

Butterfly effects in one system can interact with those in other systems to produce more complex chaotic effects. As an example, the Earth has numerous, complex, interacting, chaotic systems in which butterfly effects are continually occurring. Major ones happen in the Earth's inner and outer cores with hot electrically charged plumes of molten rock and metals rising into the mantle. This, in turn, affects the Earth's crust and this then causes continental drift and volcanic eruptions. More butterfly effects occur due to changes in ocean currents, the saltiness of the water and air temperature. The reflectivity of the Earth's surface, cloud cover, and precipitation also affect our planet's climate as does the Moon's changing gravity effects and the constantly changing angle and direction of the Earth's axis. In the biosphere, there are many changes that occur due to changes in vegetation caused by changes in the weather, insect infestation, and human development.

Butterfly effects in Earth's climate and its evolution are also caused by asteroid impacts, and cosmic rays from stars exploding in deep space. Over long time spans, changes in our planet's orbit, and our solar system's orbit through the galaxy, cause huge butterfly effects. Due to all these variables, Earth has developed an incredible diversity of very complex chemistries and environments. Earth's chemistry has produced numerous types of rocks, gems, and sediments. There are also butterfly effects

occurring from ever-changing coastlines, ocean currents, volcanism, river erosion, changing habitats, lightning, volcanic eruptions, and fires. It is no wonder climate programs are unable to predicting long-term changes when they take none of these variables into account!

Chaos rules supreme throughout the universe.

Chaos' remarkably simple formula for the Butterfly Effect is a driving force behind the evolution of all chaotic systems in the universe. It is incredible to think that such a simple equation could be responsible for the universe's infinite complexity.

Evolution driven by the Butterfly Effect not only impacts on physical systems, it also effects what most people would think of as being non- physical systems - such as the sparking of new ideas and how they can affect one's beliefs, and change cultures. Random events such as a joke taken the wrong way, an unexpected death, a change in politics, a war, new inventions, or a new discovery can all cause butterfly effects that change people's lives, their beliefs, or the economy, and much more. This is important because people come up with new ideas that can radically change the future. Butterfly effects can lead to numerous positive outcomes, such as new scientific breakthroughs, wonderful new inventions, and new forms of art etc. They can also lead to negative outcomes such as conflict, declining standards of living and education, or a collapse in law and order.

When sand grains pour chaotically into a pile, they cause random micro- avalanches at the top of the pile. But these chaotic avalanches even out to maintain a very ordered slope of 45°.

While most systems act chaotically, the overall end result is one of order. Chaotic events on one scale can create order on a larger scale. This process occurs across the universe at all levels, such as in atomic nuclei, in cells, in large biological systems, in civilizations, in economies, in solar systems, and in galaxies as well as the universe at large. **As one part of a large chaotic system decays, another part emerges to take its place, thereby maintaining overall order within the system.** This is demonstrated through endless wars. Millions of people may be killed leading up to the rise of a victor, however, sooner or later, a new war will commence, and that victor will eventually succumb to a new victor, thereby allowing the chaotic cycle to continue. We have seen this with the rise and fall of many great empires throughout history. This chaotic cycle occurs with all species. Changes in environmental conditions can cause certain species to become extinct, whilst others thrive and new species evolve, thereby maintaining order overall. Chaos also affects solar systems. Planets can have very chaotic orbits at the outset with some colliding and others being ejected out of their solar system after close encounters with other planets. The orbits of planets are continually changing over long time spans in line with Chaos Theory. The tilt of their axes can flip chaotically. But over time, chaos diminishes only to return at a later stage.

It is impossible to make accurate, long-term predictions because everything in the universe at every scale can ultimately affect everything else at every scale. We can predict however, that everything living or non-living like a star, will eventually decay. Attempting to pinpoint exactly when this will happen is unpredictable due to the Butterfly Effect.

Since Isaac Newton's time, it was thought that we had a clockwork universe. If we could get enough detailed information about it, then it would be possible to predict its future - that was until quantum mechanics was discovered in the early 20th C. It showed that there was a fundamental, inbuilt unpredictability at the sub-atomic level that makes it impossible to determine anything precisely at that level. But quantum uncertainties do not appear to be transferred to our macro universe, for reasons we have not yet identified. In the next 2-3 decades, there will be huge butterfly effects for us and all life on Earth due to the exponential advances now occurring in intelligent-computing, robotization, and genetic engineering; to mention but a few of the many rapidly progressing modern-day technologies.

Albert Einstein believed that the universe has law and order, but the great quantum physicist Neils Bohr thought that Einstein's deterministic laws were incompatible with the laws of chance at the quantum level. When Chaos Theory's Butterfly Effect was discovered, it became apparent that the universe has rigid physical laws but they work hand in hand with endless change produced by chaos. It is these two things that make our universe evolve in an ordered way at large scales but chaotically at smaller scale.

It now appears that the universe works like a gigantic blender that keeps changing environmental conditions right across the universe to force it to evolve. Without chaos, the universe would almost certainly have been stillborn.

CHAPTER 13

THE FILAMENTARY STRUCTURE OF THE UNIVERSE

When astronomers recently mapped the locations and distances of hundreds of thousands of individual galaxies, and galaxy clusters, they found that they were not evenly spaced as expected. They formed long chains in every direction. This was an amazing unforeseen discovery that was unimaginable even as little as 25 years ago. Major new discoveries like this never cease to amaze us.

From a distance, these chains look like filaments. Where filaments meet, clusters of galaxies form because this is where gravity is more concentrated. Where many filaments come together, gravity is at its greatest, so this is where superclusters are found.

The web-like filamentary structure is similar to froth. The filaments form structures like the membranes of bubbles in froth. Inside the filaments are bubble-like voids that have different sizes. Like bubbles in froth, they are all connected to those around them. But the 'bubbles' in the universe's filamentary structure are not round. The voids have few galaxies in them with none at their centers. Astronomers wonder why this is so.

The filamentary structure of galaxies across the universe and the voids between them appear somewhat similar in structure to a close-up view of fresh surf froth seen after a large wind storm. It has bubbles of many sizes all joined together.

This HST deep-field image shows chains of galaxy clusters at different distances across the universe. These form the filamentary structure discovered throughout the universe.

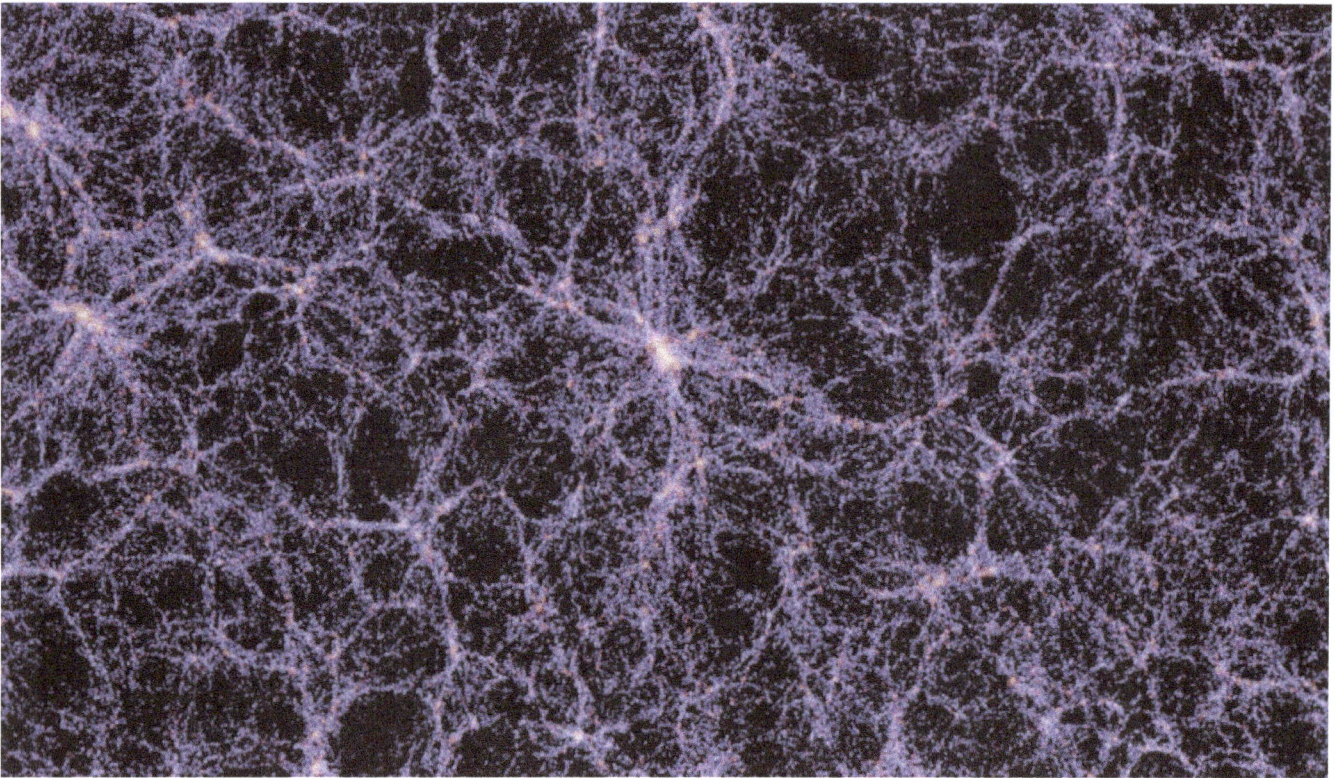

This is a 3D computer plot of the location of hundreds of thousands of galaxies. It revealed a totally unexpected, three dimensional, froth-like structure that extends throughout the universe in every direction.

The expansion of the universe is thought to be occurring because anti-gravity in the form of dark energy is causing the voids to expand more quickly with time. While space is expanding inside the voids, it does not appear to be expanding where the galaxies are because gravity is holding space there together.

When seen at close range, the filamentary structure does not appear to be homogenous, however, when seen on a large scale, it is homogeneous (i.e. the same in all directions) - just as the Big Bang Theory predicted.

Galaxies gravitate towards the outsides of the bubble membranes because this is where gravity is concentrated. The center of the voids appear to be where an anti-gravity force is strongest.

Slight fluctuations in the uniformity of the background energy level at the very beginning of the Big Bang is what is thought may have caused this structure. It is seen in the microwave background, which occurred soon after the Big Bang.

THE SPOOKY ALIGNMENT OF THE POLES OF QUASARS

Astronomers have measured the orientation of the poles of supermassive black holes that reside at the centers of 93 highly luminous quasar galaxies. Most surprisingly, the axes of these black holes are not randomly orientated! Nearly all align with the filaments in which each galaxy lies. The chance of this being through random chance is less than 1%! What force could possibly align the spin axes of supermassive black holes across the universe when all galaxies seem to be randomly orientated? Could the gravitational fields along the filaments somehow align the black holes' spin axes? Any that don't look aligned could be due to a filament running at right angles to the other filaments in front of, or behind it. Incredibly, the alignments extend for over billions of light years making them the largest known structure in the universe. Because it has taken billions of years for their light to reach us, we are seeing them as they were when the universe was only around a third to a quarter of its current age.

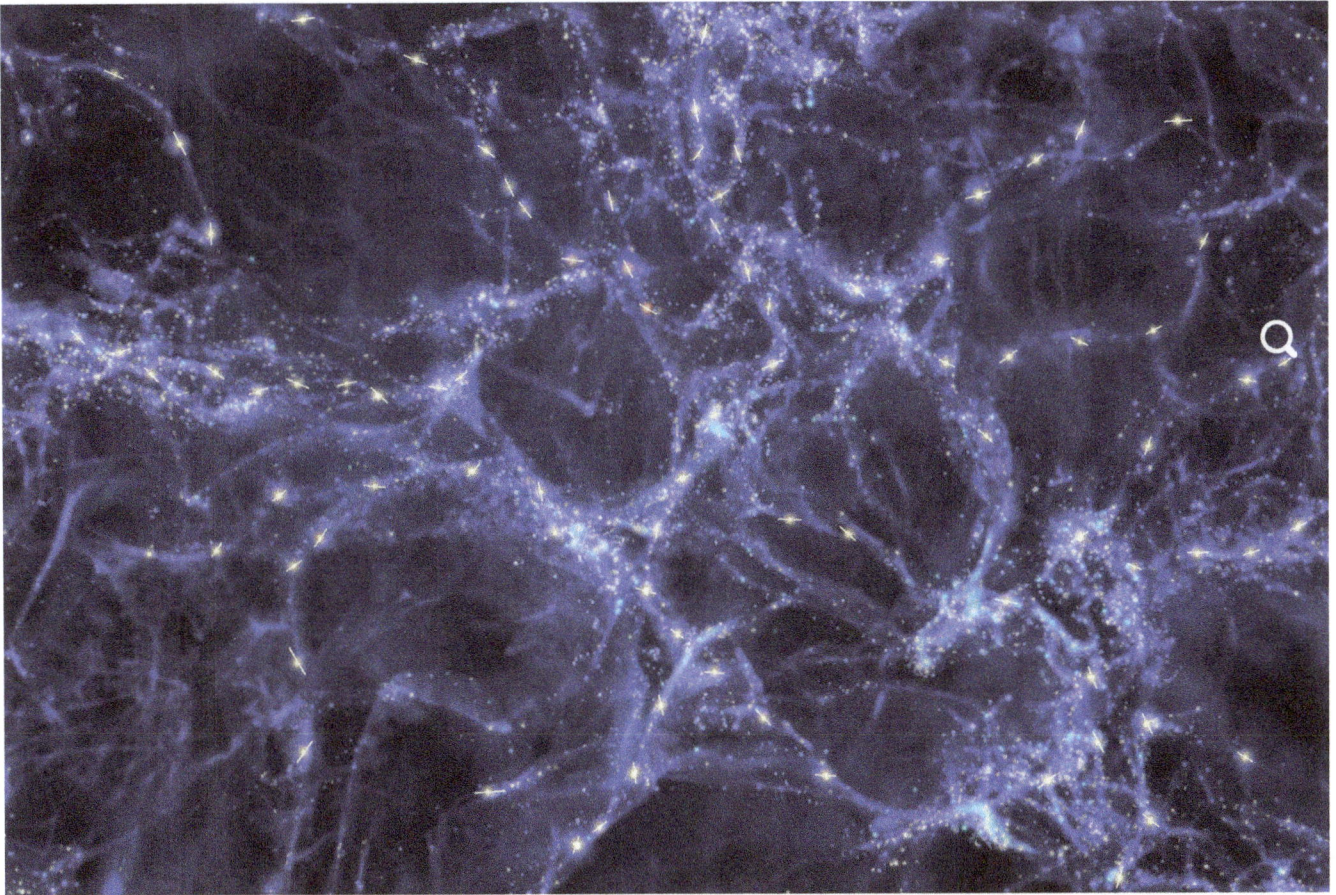

This is an illustration of the inexplicable alignment of the spin axes of supermassive black holes that lie at the centers of very active quasar galaxies that emit huge amounts of electromagnetic energy, particularly in light and radio waves. The rotation axes of the black holes are shown by a small line extending through each quasar. Illustration: ESO VLA telescope, Kommesser.

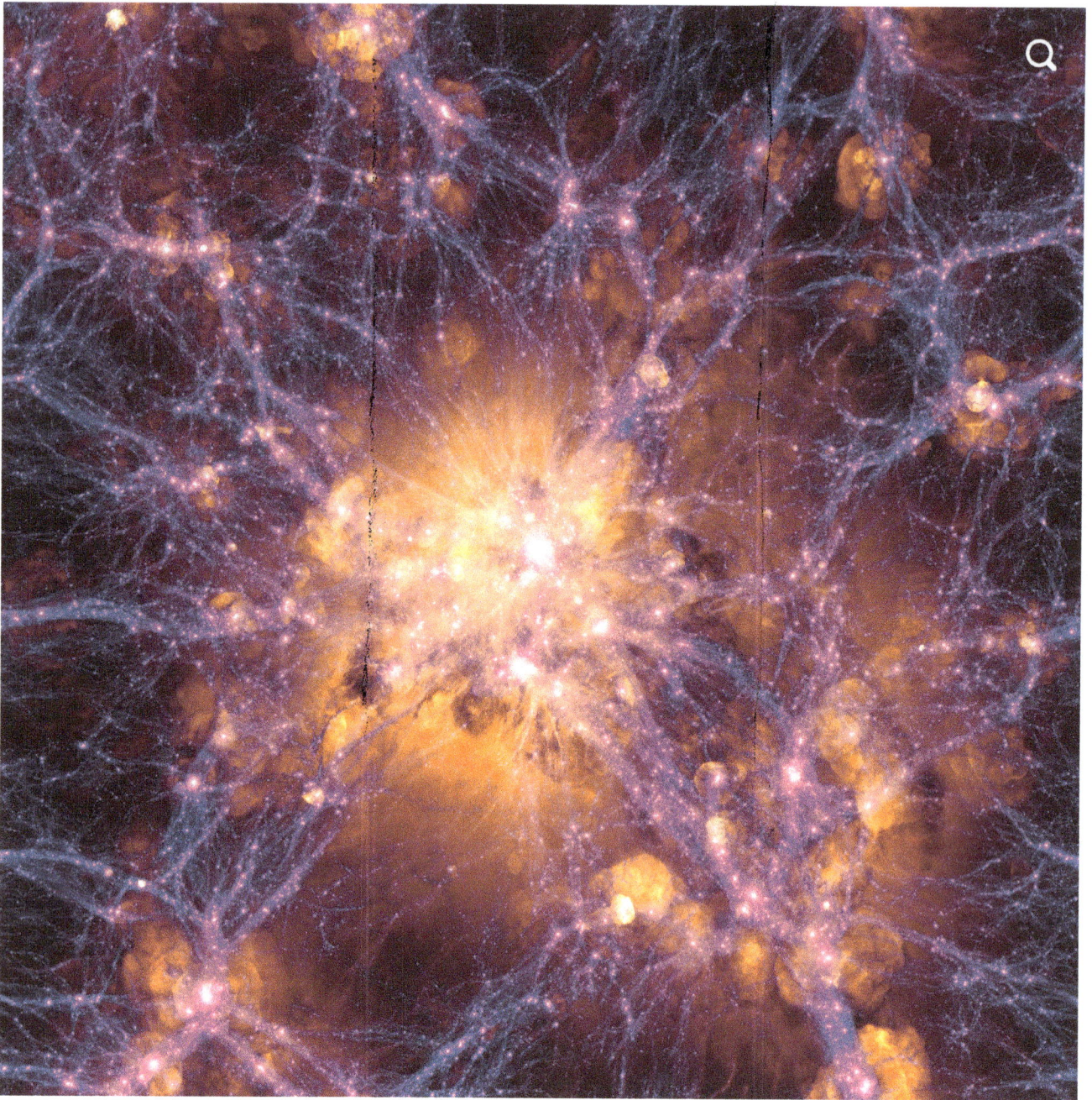

This is a single frame of an incredibly detailed video simulation of the filamentary structure of galaxies and the evolution of the universe. At the center, we see a super-massive supercluster.

The diameter of this region shown is about 300 million light years. It was created as part of the Illustris super-computer simulation. The distribution of dark matter is shown as dark blue filaments. Individual galaxies are shown as pink star-like dots. Clusters of galaxies appear white. Orange depicts the distribution of gas around the hubs of galaxy clusters. A supermassive galaxy cluster lies at the center. Supermassive black holes eject gas from their poles back into intergalactic space. This simulation is for the current age of the universe. This video is available on the Internet. Credit: ESO

THE ILLUSTRIS SIMULATION OF THE UNIVERSE

There is a stunning video of the above *Illustris simulation* available on YouTube. It was produced using the world's fastest supercomputers available at the time.

It is the most detailed simulation of the evolution of the universe ever attempted. It starts just after the Big Bang. The camera moves around the universe allowing us to see how the universe evolves. It depicts the first galaxies forming and the filamentary fabric of the universe emerging. A useful timeline of the universe's evolution is given. Enormous gaseous bubbles explode from around the centers of large clusters of galaxies. These bubbles are due to the poles of supermassive black holes ejecting gas into intergalactic space. The vision also shows a comparison between dark matter and gas temperature in a cube in the video that measures 35 million light years across. The video then zooms into the centers of galaxy clusters to reveal amazing detail. For anyone who wants to know what the big picture of the universe looks like, this video is a must. It is one of the greatest astronomical simulations you will see, especially when seen on a large, high-resolution monitor, or in an OMNIMAX theatre.

THE MICROWAVE BACKGROUND RADIATION

The filamentary structure of galaxies appears to be evenly distributed across the universe with no imbalance in any direction. This is a consequence of the uniformity revealed in the **Cosmic Microwave Background Radiation** map of the universe. It shows the minutest variations in the leftover heat from the Big Bang. This background heat is a frigid -270.15°C – a mere 3°Kelvin above absolute zero. The radiation we are seeing today has taken 13.7 billion years to arrive at our location in the universe. It has traveled from the edge of the observable universe where the Big Bang occurred at the beginning of time.

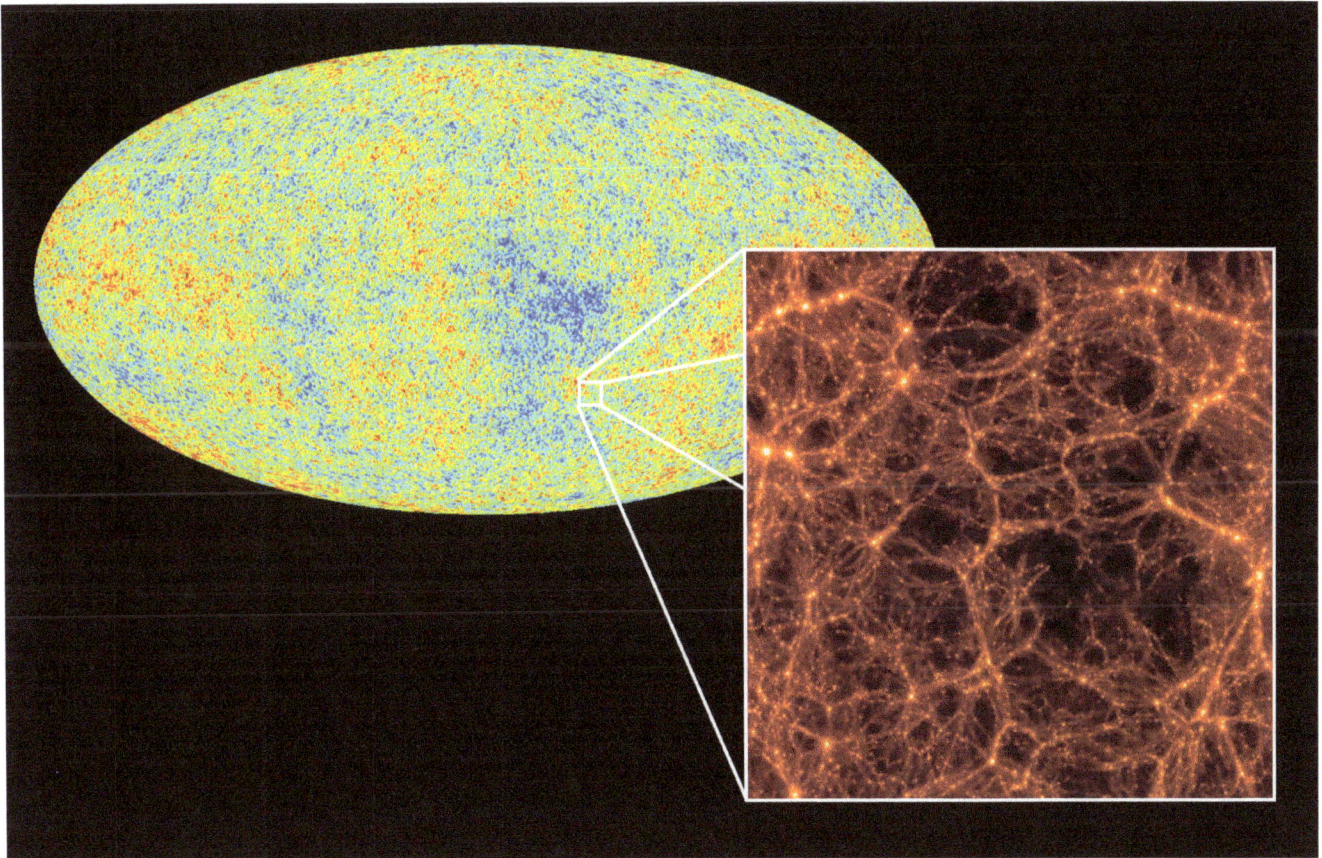

This 3°K Cosmic Microwave Background (CMB) radiation occurred one billionth of a billionth of a second after the Big Bang occurred. Because this image is so clear, it literally brought tears to the eyes of astronomers because it told them so much about the birth of the universe. The WMAP satellite that later mapped the entire sky, used even more ultra-sensitive infrared (heat) detectors. It confirmed the same result in more detail. Credit: NASA, CMB

This oval map is a full-sky picture that shows extremely small temperature variations of only 0.0002°K! These appear as variations in color. The slightly warmer regions are shown as red. This is where dark matter and normal matter became concentrated. This is the reason why superclusters of galaxies formed here. Yellow-green regions are where most galaxies originally formed before they were drawn into clusters. The slightly cooler regions are dark blue. These are voids, presumably of anti-gravity dark energy where very few galaxies exist. The square at right illustrates an expanded view of a small part of the CMB as seen in visible light. It shows the locations of the galaxies we see today, and it shows the filamentary structure that arose out of the minute temperature variations at the commencement of the universe.

CHAPTER 14

HOW DID OUR UNIVERSE BECOME SO FINELY-TUNED?

Extraordinary degrees of fine-tuning have been discovered in at least 10 physical constants in cosmology, and another 20 in quantum physics. This inexplicable degree of fine-tuning in 30 constants is infinitely beyond chance, and it's impossible to explain! There may be more constants that we have not yet discovered. This is simply mind-boggling to the nth degree.

This fine-tuning is known as the **Goldilocks enigma**. Without this extraordinary level of fine-tuning, our universe would not have become increasingly more complex, in fact, it would not have evolved at all. Over the last century, many leading cosmologists and quantum physicists have written numerous scientific papers and books on this subject trying to unravel how this could have come about, but so far, there are no convincing answers to this enigma. But this is not a reason to resort to a creator being.

There are many conditions in the universe that are precisely fine- tuned, not only for the evolution of stars and planets, but also for life to evolve. Following are just a few relatively-easy-to-grasp examples of this incomprehensible level of fine-tuning of some of the universe's constants.

1. **The four forces of nature - gravity, electromagnetism, and the strong and weak nuclear forces** - are extremely finely tuned. Even the minutest change to any one of these fundamental forces would have stopped the universe from evolving as it has, or at all.

2. **The expansion of the Big Bang** was impeccably finely-tuned. If the expansion had been just a little slower, the universe would have quickly collapsed into a big crunch and disappeared before life could have arisen. Similarly, had it exploded even the tiniest bit faster, there would not have been time for stars, planets, and life to evolve before matter had spread out too thinly. **Inexplicably, it has *precisely* the right rate of expansion!**

3. **If there was ever so *slightly* more dark energy, space-time would have expanded too fast for galaxies to form.** To have the universe we have today, the degree of dark energy had to be fantastically finely-tuned to more than 120 decimal places!

4. **If the mass and other properties of neutrons, protons, and electrons were even *infinitesimally* different, there would be no atoms in the universe.** Neutrons were created to be ever so slightly heavier than a proton so that they could decay into protons, electrons, and neutrinos. If neutrons were even the *slightest fraction* lighter, **neutrons would not decay.** Alternatively, if they were ever so slightly heavier, there would not be an abundance of hydrogen in the universe to form stars and there would be far too much helium. **Similarly, if the charge of an electron was just the tiniest bit different, there would be no complex chemistry, and therefore no life.**

5. **If the minor quantum fluctuations during the period of inflation at the outset of the Big Bang had been just minutely less, then stars and galaxies would not have formed.** Had the fluctuations been just a little stronger, then giant black holes would have formed and there would be few stars.

The leading cosmologist **Paul Davies**, author of '*The Goldilocks Enigma*' said, "*In all these parameters, the universe is balanced on an unbelievably fine knife edge.*" Many leading physicists such as **Robert Dicke, Fred Hoyle, John Gribbin** and **Martin Rees** believe that this degree of fine- tuning is far beyond the laws of chance. **Stephen Hawking** stated that "*the values of these numbers seem to have been very finely-tuned to make possible the development of life.*" These scientists are not suggesting that this was a deliberate intent by a creator god: however, it raises the question as to how this incredible degree of fine-tuning has come about? It looks like the parameters for the program that controls the universe's evolution were designed to have these precise values.

Physicist **Victor Stenger** thinks that the fine-tuning enigma may be solved if there is a unified field in which there may be connections between all the physical constants. He surmises that if a change were to occur in one constant, then the 'connections' might cause a change in one or more of the other constants thereby 'adjusting' the parameters of the universe so it can exist. No explanation is given for what the connections might be or how they would work. This is pure speculation. There is no evidence to indicate that Stenger's explanation is possible, although it is an interesting concept.

Some people have attempted to explain the fine-tuning by suggesting that a supernatural god, or some advanced alien from some super- universe outside of ours created our universe. It could equally be said that some super-mega- intelligent computer might have calculated what the conditions would need to be to create a universe like ours and then programmed this level of fine-tuning for when it turned on the Go switch. There is a problem with introducing any type of supernatural entity from outside our universe because it just raises the question as to what created the supernatural entity and its universe.

There have been many dilemmas similar to the fine-tuning enigma throughout the history of science, but they were eventually solved when new data was gained and new insights into nature were discovered. Once quantum physicists gain a much deeper understanding of the forces that underlie the quantum universe, it is possible that this may solve the fine- tuning quandary.

Leading computer technologists are now in the process of creating highly advanced, mega-intelligent, self-learning computers that will have intellects immeasurably greater than the entire human race. They may be able to gather unimaginable amounts of big data to allow them to discover how the universe became so finely tuned. If a solution to the Goldilocks' enigma is discovered, it will surely have far-reaching implications.

"The more we learn, the more complex everything in the universe becomes. The universe is not illogical or haphazard. It is an extremely finely-tuned program".

Gregg Thompson

CHAPTER 15

DID A GOD CREATE THE UNIVERSE?

As this Part challenges religious beliefs, those who are strongly religious, are advised to pass over the next section.

If a god did create the universe, which god of the thousands that humankind has worshiped would it be? Evolutionary biologist **Richard Dawkins** correctly points out in his very popular best selling book, '*The God Delusion*" that those who believe in any one of the world's 4,500 Religions and just as many cults, are atheists about every god except their own. Dawkins says that religious people only have to be atheistic about one more god (their own) to be realistic. Religion is based on illogical stories invented by persuasive men over many tens of thousands of years right through to recent times.

HOW RELIGION EVOLVED

By observing how modern-day religions and cults have sprung up, it is not hard to visualize how ancient religions preaching all-powerful, ethereal gods came into being. The god theory most likely started something like this. A clever shaman in a very primitive tribe came up with the original god concept in an attempt to explain things like life, death, emotions, disease, war, and natural disasters, etc. He would have been the first philosopher/scientist that developed a concept to explain the world. In doing so, he found that most of his tribe thought he had great insight and that his knowledge was given to him by his all-powerful god. This allowed him to have much power over his tribal members, including the chief.

This god story spread like wildfire to other tribes. Other smart opportunists in other regions realized what a big opportunity this story was to allow them to have power over their tribe so they promoted it amongst their own people. The god story was a big hit when promoted by good salesmen and convincing actors. Convincing shamans who became oracles used the god story as a pathway to gain power and wealth. They were given privileges and a life of comfort by powerful leaders.

As the god story spread, it was replicated and varied by each shaman to suit themselves and their culture. Biblical historians have shown that nearly all Catholic Popes changed biblical scripture to suit themselves. They certainly did not follow the laws they preached because secretly they did not believe in a god because they made the story up. Overwhelming evidence shows that this process is just as true today for other religions and those who start modern religions or cults.

To manipulate their believers, shamans used showmanship. They would dress in scary costumes, speak in tongues, and go into trances to look like they were in contact with their supreme being who was giving them directives for their believers to follow – directives that typically benefited the shaman of course. Remarkably, this simple trickery still works today on those who are often desperate to find something they can believe in that makes them feel like their life has meaning and is worth living, or that it at least helps them cope with their very difficult life or personality. The major churches have consolidated their power and influence through their traditions and theatrics that they use in their huge churches built to impress their brethren. They use colorful flamboyant costumes, and symbolic rituals. They have the power to influence political outcomes by guiding their gullible followers at the ballot box.

Today, charismatic preachers use sing-alongs, hypnotism, and simple special effects and illusions to give their followers a false sense that they have supernatural powers bestowed upon him by their god. People flock to them in their typically cheap, warehouse-like churches on industrial estates to sing and clap meaning into their lives to feel they are special and loved by their god. When a person wants to believe in something that makes them feel good, they dispel all realities that go against their belief in order to retain their sense of purpose. This is understandable. Unfortunately, there are many people who have so little love, respect, or joy in their life that they desperately need this. To survive, sadly, they have to hang onto any belief that deadens the physical and mental pain that reality causes them to suffer. This is grist for the mill for the opportunists who run churches and use it to manipulate them. There are however, some religious groups that do not exploit the weak but offer true pastoral care - meaningful love, companionship,

advice, and practical ways of dealing with being dealt a bad hand in life.

As large churches became established millennia ago, their powerful institutions took maximum advantage to control those under them. In a majority of civilizations, the hierarchy of the Church became the government, so they had ultimate power. This made the high priests, and the kings and those around them, very rich and authoritative. Their wealth and power assured the survival of their god myths.

In the 16th C, the Ottoman or Turkish Empire, which stretched from the Persian Gulf to Hungary, reached its peak becoming the most advanced civilization in the world. It was well ahead of all other empires in science, math, geometry, architecture, shipbuilding, strategies for war, and all of the arts, so it flourished until the Islamic religious leaders took control around 700 years ago. They realized that the practical explanations underpinning science would undermine their religious doctrine and their hold on power, so they suppressed scientific development. Those that tried to continue doing science were typically put to death. This quelling of scientific development, new scientific discoveries, and engineering breakthroughs eroded the power of their civilization. This has remained a barrier to their progress and their integration into the developed world right up to today. While most countries in the developed western world have made enormous progress scientifically, Islamic nations have not. If it was not for their luck of western companies finding oil on their lands, they would all be third world nations, struggling to survive in an inhospitable landscape.

Until recently, Christian cosmology has preached that its god was the creator of our world and all life, and that it created it all within 6 days! But the Bible never says how all that immense detail was achieved. We are expected to believe it was the greatest of all magic tricks! But recently the Catholic Church finally accepted the insurmountable evidence that our world and the universe was created by natural evolution and that it was created over 14 billion years. Does this mean that their god deceived them about how Earth was created and how long it took? Or is it more honest to say that the hierarchy of the Church knew all along that there was no god and that they did not know how the world was created? The pretense that they were in touch with a higher universal power was clearly a ruse. In the late 20th C, the Catholic Church finally admitted that they were wrong to have killed, imprisoned, and punished the world's greatest scientists hundreds of years ago for proving that the Earth went around the Sun and that Earth was not at the center of the universe as the Church preached. The Church also had to accept that the Sun was not a god as they had preached for millennia. They retained it symbolically as a halo.

Over time, some shamans developed the concept that the Sun, the Moon, and the planets, as well as some groups of stars, were gods that determined their future. This myth spread far and wide and became known as astrology.

THE ROLE OF SCIENCE AND ITS IMPACT ON HUMANITY

If one were to speculate, **it is looking increasingly more likely that the universe is a set of mathematical formulas designed to create ever higher levels of complexity.** But for what reason?

If the reason for the universe's formation turns out to be the need for it to develop increasing complexity so it can evolve ever-higher intelligence and consciousness, then we must ask, "Why does the universe seek this?" If the universe is the ultimate program, what is its end goal? I suspect that any of our present, simple, human answers to these questions would be as naive as those of a 5-year-old child trying to work out what their life is about. At the same time, they do not question how Santa Claus can visit so many homes across the world and carry presents for millions of children in his small sleigh - and do it all in one night!

Science has provided many benefits in the form of engineering solutions. This has delivered great architecture, better transportation, amazing advances in communication, irrigation, piped water, sanitation, an abundance of food, new sources of energy, and more effective weapons. But because these benefits are mostly developed in western nations, Islamic nations have to buy their technology in, and they have to send their scientists and engineers to be educated in western universities. Their scientists have been strictly controlled to mostly weapons development to prevent them from progressing in the sciences because this will challenge the Islamic religion.

As science progressed, the most advanced scientific thinkers worked out that the Sun and the Moon were not gods, but bodies out in space that moved around. By the 16th C astronomers had calculated that the planets orbited the Sun. In the 17th C, the first cells were discovered by **Robert Hooke** using a primitive microscope. It soon became apparent that cells played an enormous role in the evolution of life and people's health.

When the **Renaissance** occurred in Europe in the 17th and 18th centuries, it introduced a wave of scientific investigation known as the 'Age of Enlightenment'. This culminated in a separation of church and state. This allowed science to flourish. Once the church no longer controlled scientific inquiry, the rate of progress of scientific knowledge exploded as all topics came off the taboo list.

By the 19th C, **Charles Darwin** showed that all life forms evolved from simple cells and that the diversity of life was accomplished by natural selection. The realization that evolution acts without any need for godly intervention to create all forms of life drove a stake into the heart of most religions that preached that their god created all life forms. Christian-based religions illogically believed that the beginning of time was only 4,000 years ago, and that all life back then was the same as it is today and that all the universe and life was created by an omnipotent god in 6 days. Darwin was himself a committed Christian and the son of a preacher and thus he faced a serious personal dilemma and crisis of faith as his thesis dispelled the most fundamental teachings of his church. It took him several years to publish his book and he agonized over the impact it would have on religious belief and practice, and on his family. He was married to a deeply religious woman. All of this illustrates even further the power and sincerity of his work. When his Theory of Evolution was published, this caused him to suffer enormous personal consequences because he was virtually shunned.

Despite what religions preach has been soundly disproven, they do not die out because they offered their believers false, simple, feel-good concepts to make them feel secure and loved. Religion offers a paradigm shift and it cleverly leverages off people's insecurity by giving them false hope and they save gullible people the trouble of learning how to think for themselves. Churches tell their brethren what to do and what to think throughout their lives. Many hierarchy in the mainstream churches are now being exposed as hypocrites, frauds, unempathetic psychopaths, grand scale thieves, drug mafia, pedophiles, people smugglers and worshipers of Lucifer. This is very hard for both religious and non-religious people to come to grips with.

The best religion of all is to believe in oneself, but as this is largely controlled by one's genes, we have little say in what we believe.

IS RELIGION A COMMON STAGE OF DEVELOPMENT FOR ALL INTELLIGENCES?

In the context of the evolution of intelligent life in the universe, one has to wonder whether religion might be a fundamental evolutionary phase for all intelligent life.

At the outset of tribalism and the beginning of civilization, religion aided in developing an organized society. It provided a sense of community for many people, a value system for conducting one's life, and an explanation for one's existence. The power of the church was able to create the first forms of government. The church's system of forced donations became government taxes. It seems like everything has a purpose for a period of time.

The rise of religion as a unifying force for social order may be common for most emerging organic intelligences in the universe.

PEOPLE WHO BELIEVE IN RELIGION

Many of those who believe in a god do so because they do not want to believe that they, and their loved ones, do not exist after death. This is brilliantly illustrated in the 2009 movie '**The Invention of Lying**' where the protagonist tell his dying mother that instead of an eternity of nothingness, death brings a joyful afterlife thereby introducing the concept of a Heaven to her, so she can die happy. Word of

this concept spreads like wildfire. In the movie, believers preach this concept and he becomes the new messiah.

This appealing belief in a heaven is held by billions today despite there being not a shred of evidence for it. There is however, overwhelming evidence against it.

Religious followers want a feel-good belief system that gives them a feeling of security. They want a super being that gives meaning to their life and who they can prey to for a better life. And they want the promise of an afterlife. In contrast, most scientists do not need this: they want to discover how nature works. They are not afraid to ask difficult questions to discover how the human mind works. Intelligent, scientific investigative research to discover the truth through science has revealed much about ourselves, about life in general, and about the universe at large. None of the answers we have discovered require a god.

Some religious people think that because science has not discovered every answer to every aspect of nature, this is a reason to believe in a god. They think that their "god must exist to fill in the gaps". The few 'gaps' that remain in present scientific knowledge are quickly being filled by new discoveries. Extraordinary claims made by religious fanatics require extraordinary evidence: not supposition, and speculation, or wishful thinking. The reality is that there is not even the flimsiest evidence for the need for a god to explain the wonder of our universe.

If scientific thinking did not evolve, we would know nothing, because religion has never discovered scientific truths, but it did discover how to lie to control people and to make great wealth. In contrast, science has greatly bettered the lives of humans and many animals. Religion has led to the most unjust punishments and to innumerable wars driven by religious megalomaniacs in order for them to consolidate their power over those who are not prepared to think and learn for themselves.

"Science promotes questions, some of which may take a long time to be answered, whereas religion promotes answers that can never be questioned."
Author Unknown

The control of large populations by way of religion has been so successful that it has survived through the millennia in nearly all cultures despite the advancement of science and the use of common sense. But the power of religion is now starting to diminish. In just the last couple of decades in western countries, leftist governments have taken much power away from religious leaders to erode their ability to influence their believers, thereby allowing them to control their population.

The ethical and moral base of Christian religion that teaches us not to kill, steal, lie, or engage in pedophilia etc, is good, however, hypocritically, the hierarchy of many churches and their preachers have adeptly practiced these things, proving that the god they preach about is false. Natural moral values don't require a religion: they are common to most societies, including those that do not believe in a god. Churches and cults are largely unethical criminal businesses that pay no tax, and which are often involved in the highest financial and moral crimes. Despite many crimes being exposed in all types of religious organizations, their believers dismiss the evidence of their criminality. They have blind faith that requires no evidence to justify their belief.

If gods existed, they would be unempathetic to create so much physical pain, fear, sadness and anguish in all life forms. The inventors of most gods, including Yahweh and Allah, were promoted to be jealous of all other gods. The Koran and the First Testament of the Bible state that those who worship another god must be severely punished or killed. Having them jealous of other gods was an admission that they were not the only god and that they must have human emotions. Jealousy and murder are not the actions that one would expect from an all-knowing omnipotent gods: these are human emotions which were infused into the gods that humans invented. The concept of emotive gods wanting to kill those who do not believe in them was developed by early shamans in an attempt to stop other shamans stealing their followers.

Many psychological tests have shown that the more educated one becomes, especially in science, and the more intelligent people are, the less likely they are to believe in religion, occult beliefs, or the lies governments and the Elites tell them. This is because intelligent, knowledgeable people have more options to choose what is real and what is not. Neuroscience studies have shown that people who have genes that predispose them towards having a dominate right brain are more likely to have religious beliefs that do not require evidence. Even when provided with strong evidence that a god is extremely unlikely to exist and is not all-knowing, their genes that are driven by emotive decision-making will not let them change their mind. Until genetic engineering advances, there is nothing that can be done about this.

It's amazing that most religious people are unable to rationalize any of the points in this chapter. They cannot see how their religious leaders have manipulated them for their own gain. It is so sad that so many have so little in their lives that they have to hold onto the vaguest hope that there is someone like a Pope, or an ethereal god, that will rescue them from their misery, fear, and/or insecurity. There are religious people who are not badly off, but who are shallow thinkers who just want a simple belief that they seldom ever question by which they can live their lives. Many followers of religions know they are unknowledgeable, so they want their preacher to do their thinking for them. There are also those who feel like they are not accepted in most social groups, so by going to religious gatherings they feel a sense of friendship and belonging - as long as they pay their dues.

Why don't religious followers question such obvious anomalies in their god belief? The faiths of the Muslims, Judaists, and the Christians all arose from one original ancient religion, as did most of the world's other faiths. If there really is a god, why has he/she withheld knowledge of the real workings of the universe and of creation?

The answer is provided by science. Neuroscience has shown that people with a strongly dominant right brain are particularly prone to succumbing to religious beliefs. They have Confirmation Bias that stops them from being able to rationally question what they are told. If a god did create the universe, then who created the god and his universe? This could go on ad-infinitum.

WHAT THE FUTURE HOLDS

It seems that the more we learn, the stranger and more complex the universe becomes. This is immensely exciting, but it is challenging us to think beyond our encrusted on old beliefs, which are mostly based on simplistic thinking and a severe lack of knowledge. We are now able to learn so much about all aspects of the universe that we cannot help but wonder if there will ever be an end to the seemingly endless discoveries that can be made. We cannot possibly imagine what we will learn even in another decade, let alone in the next 50 years. Many new insights will turn many of our current beliefs upside down. This is why astronomy, cosmology, the quantum universe, the exponential advancement of technology, and neuroscience are so enthralling to inquisitive minds.

"Learn from yesterday, live for today, and hope for tomorrow. The most important thing is to not stop questioning."
Albert Einstein

FINAL WORDS

Cosmology is the most far-reaching of all sciences. It is very thought-provoking to look at the big picture of the universe and ponder its creation and where its evolution is headed.

I hope that the subjects I have included in this Volume have expanded your Big Picture view of the cosmos. If you are a questioning person who loves to delve into the 'why' of everything, as I do, then you should have many questions that will make you want to do deep research to become aware of new discoveries as they come to light. These will expand your view of our universe from the sub-atomic level to the broad scale of the entire universe from its birth to its death. And then try to consider where we fit into its grand design.

I hope I have opened your mind to many concepts that you may not have considered.

Gregg Thompson

INDEX